Queueing theory essentials for engineers

José Alberto Hernández and Pablo Serrano
Universidad Carlos III de Madrid

Copyright © 2025
ISBN: 9798327027183

This book is dedicated to all engineers who make our lives better

Contents

1 **Probability review** 1
   1.1 Random variables 1
   1.2 Probability Density Functions 2
   1.3 Expectation and variance 3
   1.4 Some useful PDFs 5
   1.5 Conditional and joint probability 6
   1.6 Conditional expectation 8
   1.7 The Bayes' theorem 9
   1.8 The Central Limit Theorem 10
   1.9 Further problems 12

2 **The exponential random variable** 27
   2.1 Definition 27
   2.2 Simulation and estimation 28
   2.3 The memoryless property 30
   2.4 Minimum of several exponential random variables 32
   2.5 Comparison of several exponential random variables 34
   2.6 Further problems 35

3 **The Poisson process** 45
   3.1 Counting processes 45
   3.2 Stationary and independent increments 46
   3.3 The Poisson process 47
   3.4 Inter-arrival times of events 49
   3.5 Waiting times 50
   3.6 Conditional distribution of arrivals 52

    3.7 Aggregation and sampling of Poisson processes   53
    3.8 The theorem of Palm-Khintchine 54
    3.9 Further problems 56

4 Discrete-Time Markov Chains   69

    4.1 Introduction by example 69
    4.2 States and transitions between states 70
    4.3 Problem formulation 72
    4.4 The Markov property and the Chapman-Kolmogorov equations 75
    4.5 Sojourn times 77
    4.6 Reachability and types of states 78
    4.7 Steady-state analysis 80
    4.8 First-passage times 82
    4.9 Further problems 83

5 Continuous-Time Markov Chains   97

    5.1 CTMCs by example 97
    5.2 The infinitesimal generator 101
    5.3 Definition and Chapman-Kolmogorov equations 104
    5.4 The transient behavior of a CTMC 105
    5.5 Steady-state analysis and balance equations . . . 108
    5.6 First-passage times 111
    5.7 Further problems 113

6 Classical queueing theory   127

    6.1 Introduction by example 127
    6.2 Definition and Kendall's notation 130
    6.3 Little's theorem 134
    6.4 The classical M/M/1 queue 136
    6.5 The M/M/1/K queueing system 141
    6.6 The M/M/c queueing system 144
    6.7 Other non-classical Markovian queueing systems 150
    6.8 Further problems 153

7 Open queueing networks   175

    7.1 Introduction 175

7.2 Burke's theorem 176
7.3 Analysis of two M/M/1 queues in tandem . . . 178
7.4 Types of networks and Jackson's theorem 179
7.5 Extended analysis for other open networks and Kleinrock's approximation 182
7.6 Further problems 185

Index   **199**

Preface

About the authors

José Alberto Hernández and Pablo Serrano are senior researchers at the Department of Telematic Engineering of Universidad Carlos III de Madrid, Spain. In 2009, they designed a course on performance evaluation of networks for the second year of the Bachelor's Degree in Communications and Networks at the university. This book comprises more than 15 years of experience teaching this course. The feedback collected from students has been very useful in understanding which concepts require further explanation and examples for a better understanding.

José Alberto Hernández completed a Bachelor's degree in Telecommunication Engineering at Universidad Carlos III de Madrid (Spain) in 2002, and a Ph.D. degree in Computer Science at Loughborough University (Leics, United Kingdom) in 2005. From 2005 to 2009, he was a postdoctoral researcher and teaching assistant at Universidad Autónoma de Madrid, where he participated in a number of both national and European research projects concerning the modeling and performance evaluation of communication networks. In 2009, he moved to Universidad Carlos III de Madrid, where he became an Associate Professor. His research interests include the areas in which mathematical modeling and computer networks overlap. He has published more than 150 scientific articles in both peer-reviewed journals and conferences in proceedings, in the areas of network performance and reliability, broadband networks, high-speed switched networks, optical WDM, converged optical-wireless access networks, and energy efficiency.

Pablo Serrano obtained his Telecommunication Engineering degree and his PhD from Universidad Carlos III de Madrid (UC3M) in 2002 and 2006, respectively. He has been with the Telematics Department of UC3M since 2002, where he currently holds the position of Associate Professor. In 2007 he was a Visiting Researcher at the Computer Network Research Group at Univ. of Massachusetts Amherst partially supported by the Spanish Ministry of Education under a José Castillejo grant. He has published over 150 scientific papers in peer-reviewed international journals and conferences. He serves on the Editorial Board of IEEE Communications Letters and has served on the TPC of several conferences and workshops including IEEE Globecom and IEEE INFOCOM. His current work focuses on the performance evaluation of wireless networks.

Book structure

This book is split into seven chapters, not isolated but dependent.

The first chapter provides a basic review of some important probabilistic concepts and theorems, including expectation, joint and conditional probability, Bayes' theorem, etc. Some well-known Probability Density Functions are also introduced, and most of them will be thoroughly used across the book.

The second and third chapters focus on the exponential distribution and the Poisson process respectively. The *memoryless property of the exponential distribution* comprises the first important milestone of the book since it will play a key role in the Poisson process, and continuous-time Markov chains. We often say in class that the exponential distribution and

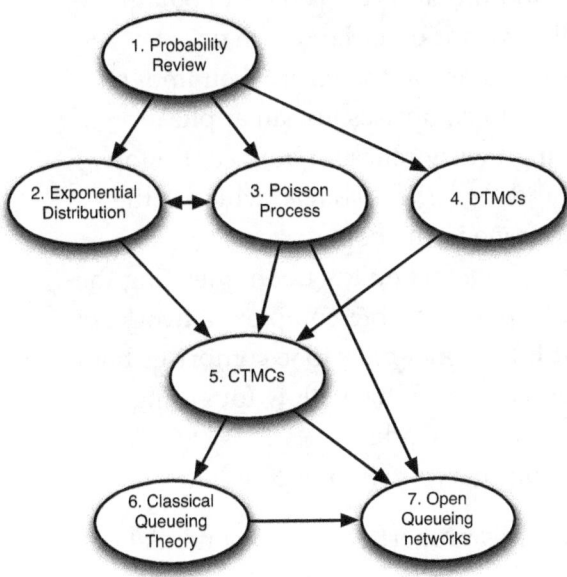

Figure 1: Book contents and relationships between chapters.

the Poisson process are like *husband and wife* since they go everywhere together.

Then, the book introduces *discrete-time Markov chains* which are very useful in the analysis of a large number of communication systems, protocols, and network-related situations. The *Chapman-Kolmogorov equations* and the concept of *limiting-state probabilities* comprise the second big milestone of this book.

The next step is to present *continuous-time Markov chains* which combine discrete-time Markov chains with exponential distributions. Before addressing this chapter, the properties of the exponential distribution and the analysis of discrete-time Markov chains must be very clear to the reader.

The third milestone in the book comprises the ability to correctly use a continuous-time Markov chain to model a given problem, including the *state-transition-rate diagram*, the *infinitesimal generator* and the *balance equations* to obtain the steady-state probability vector of a Markov chain.

Once continuous-time Markov chains are well understood, then *queueing theory* comprises just an application of them. Indeed, queueing theory relies on both continuous-time Markov chains and *the Little's theorem*, which is the fourth important milestone in the book.

Finally, *open queueing networks* builds upon queueing theory and provides tools to analyze big complex networks of queues easily. The fifth and final milestone comprises both the *Burke's and Jackson's theorem* which allow for easing the analysis of open queueing networks.

The following is a summary of all milestones:

- *Milestone 1:* Memoryless property of the exponential distribution.

- *Milestone 2:* Chapman-Kolmogorov equations and limiting-state probabilities of discrete-time Markov chains.

- *Milestone 3:* State-transition-rate diagram, infinitesimal generator, and balance equations of continuous-time Markov chains.

- *Milestone 4:* Little's theorem and Kendall's notation.

- *Milestone 5:* Burke's and Jackson's theorems.

Focus and readership

The book aims at providing step-by-step learning, starting from intuitive examples, introducing the theory, and strengthening the learning process with new exercises. The goal is to involve the reader through the learning process with small frequent challenges in every chapter to make him/her feel confident with the theory.

The length of the book has been deliberately intended to remain short, giving priority to examples and problems over long theorem proofs. The reader is recommended to visit other classical books (those pointed out in the bibliography) for further details concerning theorems and their proofs, since we have more focused on intuition rather than rigor. The book includes over 100 solved examples and exercises that cover the main concepts introduced in theory, with an application to computer networks and communication protocols.

The contents of this book have been designed to span a whole semester in the third or fourth year of a degree in Computer Science or Electrical Engineering. In addition, this book might be suitable for students conducting their research in fields related to the performance evaluation of computer networks.

In light of this, the book assumes that the reader is already familiar with probability theory and communication networks, especially concerning layer-2 protocols including Medium Access Control (MAC) protocols and TCP/IP fundamentals. In addition, it is recommended that the author has some programming skills in Octave or Matlab.

Acknowledgements

The writers would like to acknowledge the help of Dr. Ignacio Soto, Dr. Isaías Martínez, and Dr. Andrés García for valuable comments and proof-reading.

1
Probability review

1.1 Random variables

Consider a random experiment having sample space S. In probability theory, the sample space S of an experiment or random trial is the set of all possible outcomes.

Random variable A random variable X is a variable whose value is unknown. Its possible values usually represent the possible outcomes of an experiment.

For example:

1. Tossing a coin may have two possible outcomes: $S_X = \{Heads, Tails\}$.

2. Rolling a die may have six possible outcomes: $S_X = \{1, 2, 3, 4, 5, 6\}$.

3. The height of adult human beings: $S_X \in (150, 235)$ cm.

Random variables may be continuously or discretely valued. The first two cases are examples of discrete random variables, whereas the last represents a continuous random variable.

1.2 Probability Density Functions

The Probability Density Function (PDF) describes the relative likelihood of every possible outcome of an experiment characterized by some random variable.

In the previous examples, the random variables can be characterized by the following PDFs:

1. Tossing a coin: $P(X = Heads) = P(X = Tails) = 0.5$
2. Rolling a die: $P(X = i) = \frac{1}{6}$, with $i = 1, \ldots, 6$ (uniformly distributed).
3. Human height: $f_X(x) \sim N(\mu, \sigma)$

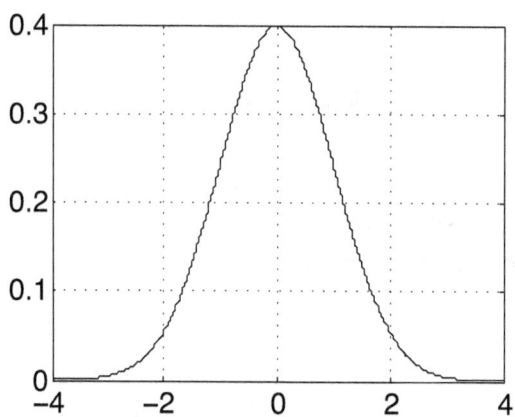

Figure 1.1: Normal PDF, N(0,1).

It is worth remarking that:
$$\sum_x P(X = x) = 1 \qquad (1.1)$$
for discrete random variables, and:
$$\int_x f_X(x)dx = 1 \qquad (1.2)$$
for continuous random variables.

The Cumulative Distribution Function (CDF) gives the probability that a random variable takes a value below or equal to x, i.e.:

$$F_X(x) = P(X \leq x) = \int_{-\infty}^{x} f_X(\tau) d\tau \quad \text{cont. r.v.} \quad (1.3)$$

$$F_X(x) = P(X \leq x) = \sum_{k \leq x} P(X = k) \quad \text{disc. r.v.} \quad (1.4)$$

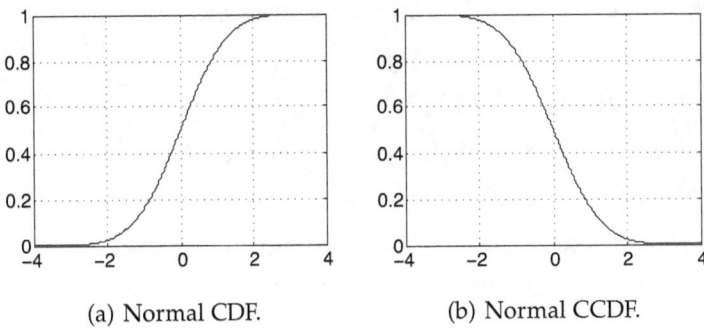

(a) Normal CDF.   (b) Normal CCDF.

Figure 1.2: CDF and CCDF of the standard Normal distribution $N(0,1)$

Finally, the Complementary Cumulative Distribution Function (CCDF) or Survival function gives the probability of taking a value greater than x:

$$S_X(x) = P(X > x) = 1 - F_X(x) \quad (1.5)$$

In other words:

$$S_X(x) = \int_{x}^{\infty} f_X(\tau) d\tau \quad \text{cont. r.v.} \quad (1.6)$$

$$S_X(x) = \sum_{k > x} P(X = k) \quad \text{disc. r.v.} \quad (1.7)$$

1.3 Expectation and variance

The expectation and variance provide a good summary of the properties of a random variable.

Expectation The expected value of a random variable is the weighted average of all possible outcomes for the

experiment it characterizes. Mathematically:
$$E(X) = \sum_x xP(X=x) \qquad (1.8)$$
for discrete random variables, and:
$$E(X) = \int_x x f_X(x) dx \qquad (1.9)$$
for continuous random variables.

It can be shown that the expectation is a linear operator:
$$E(aX_1 + bX_2 + c) =$$
$$= aE(X_1) + bE(X_2) + c$$
for any two random variables (*independent or not*) X and Y and any real numbers a, b and c.

Variance The variance is a measure of how spread the possible outcomes of an experiment are concerning its expectation.
$$Var(X) = \sum_x [x - E(X)]^2 P(X=x) \qquad (1.10)$$
for discrete random variables, and:
$$Var(X) = \int_x [x - E(X)]^2 f_X(x) dx \qquad (1.11)$$
for continuous random variables.

The variance is not a linear operator:
$$Var(aX_1 + bX_2 + c) =$$
$$= a^2 Var(X_1) + b^2 Var(X_2)$$
for any two *independent* random variables X and Y and any real numbers a, b and c.

Alternatively, the variance may be computed as:
$$Var(X) = E(X^2) - (E(X))^2 \qquad (1.12)$$
where the second moment $E(X^2)$ is:
$$E(X^2) = \sum_x x^2 P(X=x) \qquad (1.13)$$
for discrete random variables, and:
$$E(X^2) = \int_x x^2 f_X(x) dx \qquad (1.14)$$
for continuous random variables.

Example 1

Find the expectation and variance of rolling a die.

Solution

The expectation and the variance for the experiment of rolling a die are:
$$E(X) = 1\frac{1}{6} + 2\frac{1}{6} + \ldots + 6\frac{1}{6} = \frac{7}{2} = 3.5$$

Now,
$$E(X^2) = 1^2\frac{1}{6} + 2^2\frac{1}{6} + \ldots + 6^2\frac{1}{6} = \frac{91}{6}$$

which yields:
$$Var(X) = E(X^2) - (E(X))^2 = \frac{91}{6} - \left(\frac{7}{2}\right)^2 = \frac{35}{12} \approx 2.92$$

1.4 Some useful PDFs

The following list provides some useful PDFs along with their expectation and variance values:

1. Continuous Uniform $U(a,b)$:
$$f_X(x) = \frac{1}{b-a}, \quad a \leq x \leq b \qquad (1.15)$$
$$E(X) = \frac{a+b}{2}, \quad Var(X) = \frac{(b-a)^2}{12}$$

2. Discrete Uniform $U(a,b)$:
$$P(X = k_i) = \frac{1}{n}, \quad i = 1, \ldots, n \qquad (1.16)$$
$$E(X) = \frac{1}{n}\sum_{i=1}^{n} k_i, \quad Var(X) = \frac{1}{n}\sum_{i=1}^{n}(k_i - E(X))^2$$

3. Binomial $Binom(N, p)$:
$$P(X = k) = \binom{N}{k} p^k (1-p)^{N-k}, \quad k = 0, \ldots, N \qquad (1.17)$$
$$E(X) = Np, \quad Var(X) = Np(1-p)$$

4. Normal distribution $N(\mu, \sigma^2)$:

$$f_X(x) = \frac{1}{\sqrt{2\pi\sigma^2}} e^{-\frac{(x-\mu)^2}{2\sigma^2}}, \quad -\infty \leq x \leq \infty \tag{1.18}$$

$$E(X) = \mu, \quad Var(X) = \sigma^2$$

5. Exponential distribution $Exp(\lambda)$:

$$f_X(x) = \lambda e^{-\lambda t}, \quad t \geq 0 \tag{1.19}$$

$$E(X) = \frac{1}{\lambda}, \quad Var(X) = \frac{1}{\lambda^2}$$

6. Geometric distribution $Geo(p)$:

$$P(X = k) = (1-p)p^k, \quad k = 0, 1, 2, \ldots \tag{1.20}$$

$$E(X) = \frac{p}{1-p}, \quad Var(X) = \frac{p}{(1-p)^2}$$

7. Poisson distribution $Poiss(\lambda)$:

$$P(X = k) = \frac{\lambda^k}{k!} e^{-\lambda}, \quad k = 0, 1, 2, \ldots \tag{1.21}$$

$$E(X) = \lambda, \quad Var(X) = \lambda$$

1.5 Conditional and joint probability

Conditional probability refers to the probability of some event X, given the occurrence of some other event Y, and is represented as:

$$P(X|Y) = \frac{P(X,Y)}{P(Y)} \tag{1.22}$$

where $P(X, Y)$ is referred to as the *joint probability* of random variables X and Y. Such a joint probability accounts for the case when both X and Y occur.

If the random variables X and Y are independent, then:

$$P(X,Y) = P(X)P(Y) \tag{1.23}$$

The next example illustrates the concept of joint and conditional probability.

Example 2

Consider the experiment of rolling two different dice. Let D_1 and D_2 refer to the random variables that represent the result of each die separately, and $X = D_1 + D_2$ refer to the sum. Can you find the values of $P(X = 4)$ and $P(X = 4|D_1 = 1)$ and compare the results?

Solution

First, the sample space of X is $S_X = \{2, 3, \ldots, 12\}$.
There are 36 possible outcomes of the experiment (D_1, D_2), listed in the next Table:

(1,1)	(1,2)	(1,3)	(1,4)	(1,5)	(1,6)
(2,1)	(2,2)	(2,3)	(2,4)	(2,5)	(2,6)
(3,1)	(3,2)	(3,3)	(3,4)	(3,5)	(3,6)
(4,1)	(4,2)	(4,3)	(4,4)	(4,5)	(4,6)
(5,1)	(5,2)	(5,3)	(5,4)	(5,5)	(5,6)
(6,1)	(6,2)	(6,3)	(6,4)	(6,5)	(6,6)

For instance, the joint probability of having the outcome $(1, 3)$ is:

$$P(D_1 = 1, D_2 = 3) = P(D_1 = 1)P(D_2 = 3) = \frac{1}{6} \times \frac{1}{6} = \frac{1}{36}$$

since the results obtained from rolling the two dice are independent.
However, the probability of scoring $X = 4$ is:

$$P(X = 4) = \frac{3}{36}$$

since this value takes into account the three possible outcomes whose sum equals 4: $(1, 3)$, $(2, 2)$, and $(3, 1)$ (see the table above).
Next, consider the following probability value:

$$P(X = 4|D_1 = 1)$$

This value should be different than $P(X = 4)$ since some information is already provided, i.e. the result of the first die: $D_1 = 1$. Indeed:

$$P(X = 4|D_1 = 1) = \frac{1}{6} \neq P(X = 4) = \frac{3}{36}$$

Essentially, given the fact that D_1 scored a one, then D_2 has to take the value: $D_2 = 3$ to make $X = 1 + 3 = 4$, otherwise it would not be possible to score a 4. Then:

$$P(X = 4 | D_1 = 1) = P(D_2 = 3) = \frac{1}{6}$$

Essentially, the fact that $D_1 = 1$ is known makes us consider only the first column of the previous table, i.e. results of type $(1, *)$. Then, $D_2 = 3$ is just one case out of six possible.

Alternatively, the previous result can also be obtained from eq. 1.22:

$$P(X = 4 | D_1 = 1) = \frac{P(X = 4, D_1 = 1)}{P(D_1 = 1)} = \frac{1/36}{1/6} = \frac{1}{6}$$

since both conditions, $X = 4$ and $D_1 = 1$, are only met in one case out of 36 possible (the case $(1,3)$).

Example 3

Can you find the value of $P(D_1 = 1 | X = 4)$?

Solution

This example requires to consider all possible cases for $X = 4$ which are: $(1,3)$, $(2,2)$, and $(3,1)$. The value $P(D_1 = 1 | X = 4)$ requires to see how many of those outcomes happen to have $D_1 = 1$. The answer is only one: $(1,3)$ out of three possible. Thus:

$$P(D_1 = 1 | X = 4) = \frac{1}{3}$$

Alternatively, the same result arises after applying eq. 1.22:

$$P(D_1 = 1 | X = 4) = \frac{P(D_1 = 1, X = 4)}{P(X = 4)} = \frac{1/36}{3/36} = \frac{1}{3}$$

1.6 Conditional expectation

The conditional expectation is the expected value of a random variable concerning a conditional probability distribution. Mathematically:

$$E(X|Y=y) = \sum_x xP(X=x|Y=y) \quad (1.24)$$

Example 4

Can you obtain $E(X|D_1 = 1)$ for the two-dice example?

Solution

We may compute $E(X|D_1 = 1)$ as:

$$E(X|D_1 = 1) = \sum_x xP(X=x|D_1=1)$$

$$= 2\frac{1}{6} + 3\frac{1}{6} + \ldots + 7\frac{1}{6} = 4.5$$

Alternatively, the previous value can also be obtained from:

$$E(X|D_1 = 1) = 1 + E(D_2) = 1 + 3.5 = 4.5$$

since $D_1 = 1$ is known.

Finally, it can be shown that:

$$E(X) = E(E(X|Y)) = \sum_y E(X|Y=y)P(Y=y) \quad (1.25)$$

1.7 The Bayes' theorem

The following two theorems are very useful in many probabilistic problems:

Total probability theorem Given n mutually exclusive events $E_1, E_2, \ldots, E_n$ whose probabilities sum to unity, then:

$$P(X) = P(X|E_1)P(E_1) + \ldots P(X|E_n)P(E_n) \quad (1.26)$$

Bayes' theorem The Bayes' theorem shows the relationship between two conditional probabilities:

$$P(E_k|X) = \frac{P(X, E_k)}{P(X)} = \frac{P(X|E_k)P(E_k)}{\sum_i P(X|E_i)P(E_i)} \quad (1.27)$$

Example 5

Consider two urns, the first one contains two white and six black balls, and the second one contains six white and two black balls.

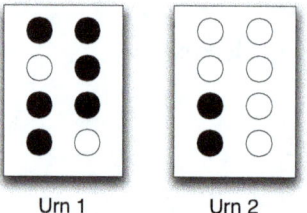

Urn 1   Urn 2

We toss a fair coin, and then we draw a ball from the first urn or the second urn, depending on whether the outcome of the coin is heads or tails. The first question is: What is the probability that a white ball is drawn? The second question is: What is the conditional probability that the outcome of the toss was heads given that a white ball was selected?

Solution

The first question requires making use of the total probability theorem:

$$P(W) = P(W|U_1)P(U_1) + P(W|U_2)P(U_2)$$
$$= \frac{2}{2+6} \times \frac{1}{2} + \frac{6}{6+2} \times \frac{1}{2} = \frac{1}{2}$$

The solution to the second question arises after applying Bayes' theorem:

$$P(U_1|W) = \frac{P(W|U_1)P(U_1)}{P(W)} = \frac{\frac{2}{2+6}\frac{1}{2}}{\frac{1}{2}} = \frac{1}{4}$$

1.8 The Central Limit Theorem

Consider the following set of random variables: $\{X_1, X_2, \ldots, X_n\}$, each of which happens to be independent and identically distributed (i.i.d.) with finite mean μ and variance σ^2. We know that the sum:

$$S_n = X_1 + \ldots + X_n$$

has mean $n\mu$ and variance $n\sigma^2$.

The Central Limit Theorem states that the random variable:

$$Z_n = \frac{S_n - n\mu}{\sigma\sqrt{n}}$$

approximates the standard normal distribution $N(0,1)$ for large n. In particular:

$$\lim_{n \to \infty} P\left(\frac{S_n - n\mu}{\sigma\sqrt{n}} \leq z\right) = \frac{1}{\sqrt{2\pi}} \int_{-\infty}^{z} e^{-\frac{u^2}{2}} du \qquad (1.28)$$

Example 6

Let S_n denote the result of rolling n dice and adding up their results. Can you find the PDF of S_n?

Solution

Thanks to the Central Limit Theorem, we know that the PDF of S_n approaches a Normal distribution $N(\frac{42}{12}n, \frac{35}{12}n)$.

The next figures below show examples of the the applicability of the CLT for the sum of two S_2 and ten S_{10} dice respectively.

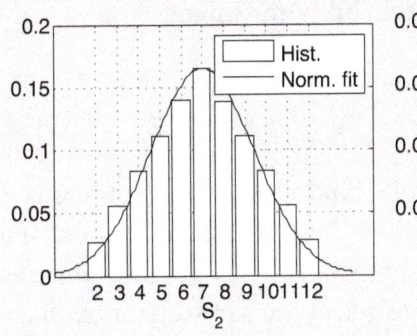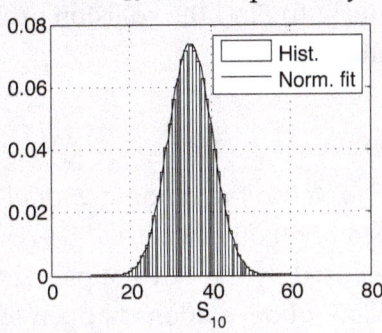

1.9 Further problems

Problem 1

Consider a computer that may send Data frames or Acknowledgements with equal probability. Compute the probability that the first time two data frames are transmitted consecutively on the third and fourth frames.

Solution

This experiment requires that the third and fourth frames are Data frames. Also, the second frame must be an Ack, otherwise, we would have the first data frames on the second and third frames instead of the third and fourth. Finally, the first frame can either be a Data frame or Ack. Hence:

$$P(*, Ack, Data, Data) = P(*)P(Ack)P(Data)P(Data)$$
$$= 1 \times 0.5 \times 0.5 \times 0.5 = 0.125$$

Problem 2

Consider a wireless network with 13 radio channels. In this network, two computers decide to use any of these channels for communicating with an access point. There is a collision if any two computers use the same channel. a) Compute the probability of having a collision when the network comprises only two computers. b) Find the collision probability for three computers.

Solution

Let CO_2 refer to the random variable that takes value 1 if there is a collision or 0 otherwise, i.e. $S_{CO_2} = \{0,1\}$; and let C_i refer to the channel used by the i-th station, where $S_{C_i} = \{1,\ldots,13\}$. There is a collision if both stations happen to select the same channel, that is:

$$\begin{aligned} P(CO_2 = 1) &= P(C_1 = 1, C_2 = 1) + P(C_1 = 2, C_2 = 2) + \ldots \\ &\quad + P(C_1 = 13, C_2 = 13) \\ &= \frac{1}{13}\frac{1}{13} + \ldots + \frac{1}{13}\frac{1}{13} = 13 \times \frac{1}{13^2} = \frac{1}{13} \end{aligned}$$

Also, this probability may be obtained by counting the number of cases that produce a collision and dividing this number by the total number of cases. There are thirteen collision cases:

$$(1,1), (2,2), \ldots, (13,13)$$

over a total number of cases of $13 \times 13 = 169$. Thus:

$$P(CO_2 = 1) = \frac{\text{no. possible cases}}{\text{no. total cases}} = \frac{13}{13^2} = \frac{1}{13}$$

Similarly, there are $13^3 = 2197$ total cases when we have three stations instead of two.

Now, we need to compute the number of collision cases. For channel no. 1, we have three sets of collision cases: $(1,1,*)$, $(1,*,1)$, and $(*,1,1)$, where $*$ may take any value from 2 to 13. That is a total of 12×3 cases. The same reasoning applies to channel 2: $(2,2,*)$, $(2,*,2)$ and $(*,2,2)$ yielding another set of 12×3 set of collision cases. The same applies for channels $3, \ldots, 13$. Thus, we have $13 \times 12 \times 3$ cases where any two stations select the same two channels. Finally, we also need to include the cases where all three stations happen to select the same channel:

$$(1,1,1), (2,2,2), \ldots, (13,13,13)$$

which is another set of 13 cases. Thus:

$$P(CO_3 = 1) = \frac{13 \times 12 \times 3 + 13}{13^3} = \frac{37}{169}$$

Alternatively, the same result arises with the following reasoning: There is no collision if the second station does not select the same channel as the first station, and the third station does not select the same channel as station 1 or 2. Hence, the first station has 13 channels to choose from, the second station only has 12 channels to choose from (one is already occupied by station 1), while the third station only has 11 channels to choose from (two channels are occupied by stations 1 and 2). Thus, the total number of non-collision cases is: $13 \times 12 \times 11 = 1716$. Hence:

$$P(CO_3 = 1) = 1 - \frac{13 \times 12 \times 11}{13^3} = \frac{37}{169}$$

Problem 3

The amount of traffic received by a computer is measured in the next table.

App.	Packet size dist.	%
Skype	$U(50,150)$	5%
P2P	$U(1000,1500)$	60%
Web	$Exp(1/1000)$	25%
email	$N(800,100)$	10%

In this table, the first column gives the application type of the packets, the second column refers to the packet size distribution for that application and the third column indicates the percentage of traffic over the total for such an application. Obtain the average packet size.

Solution

This problem requires to make use of the conditional expectation equations. Let PS refer to the random variable denoting packet size, and let A refer to the application type. Then:

$$\begin{aligned} E(PS) &= E(E(PS|A)) = \sum_A E(PS|A) P(A) \\ &= 0.05 \times \frac{150 + 50}{2} + 0.6 \times \frac{1500 + 1000}{2} \\ &\quad + 0.25 \times 1000 + 0.1 \times 800 \\ &= 1085 \text{ Bytes} \end{aligned}$$

which gives the average packet size weighted per application.

Problem 4

A wireless network uses Stop-and-Wait ARQ to recover lost frames in lossy media. Such a lossy media has a frame loss probability of 0.1 and a Maximum Transfer Unit (MTU) of 1000-byte packets. Assume, that a given user wants to download a file of 10000 bytes. Compute (a) the probability that no frame needs retransmission, and (b) exactly two frames need retransmission. Compute also (c) the average number of frames that need retransmission.

Solution

First, a retransmission is needed in Stop-and-Wait ARQ if either a data frame or its Ack (or both) are lost. This occurs with probability:

$$\begin{aligned} p_{retx} &= P(Data = Lost, Ack = Lost) \\ &+ P(Data = Ok, Ack = Lost) \\ &+ P(Data = Lost, Ack = Ok) \\ &= 0.1 \cdot 0.1 + 0.9 \cdot 0.1 + 0.1 \cdot 0.9 = 0.19 \end{aligned}$$

Also, the number of data frames to be transmitted is:

$$N = \frac{\text{file size}}{\text{MTU}} = 10$$

The number of retransmissions required can be modeled with the binomial distribution. That is, k frames require one or more retransmissions with probability:

$$P(k) = \binom{N}{k} p_{retx}^k (1 - p_{retx})^{N-k}$$

Hence, no frame needs retransmission with probability:

$$P(k = 0) = \binom{10}{0} p_{retx}^0 (1 - p_{retx})^{10-0} = (1 - 0.19)^{10} = 0.1216$$

Similarly, exactly two frames need retransmission with probability:

$$P(k = 2) = \binom{10}{2} p_{retx}^2 (1 - p_{retx})^8 = 0.301$$

The average number of retransmissions needed is:
$$E(k) = Np = 1.9 \text{ frames}$$
so about 2 frames are expected to require retransmission.

Problem 5

A server in a company has four processors (four cores) to serve users' jobs. Consider that only three users can log into the server. Each user may offer zero, one, or two jobs, with the following probabilities: he offers one job with probability $2p$, and two jobs with probability p. (a) Find the average number of used processors when $p = 0.1$. Find also the variance. (b) Find the value of p that achieves 100% utilization of the four processors on average. (c) For the value of p computed in the previous case, obtain the probability of having more jobs than processors.

Solution

First, let U_i refer to the number of jobs offered by the i-th user, with $S_{U_i} = \{0, 1, 2\}$. The total number of jobs offered is:
$$U = U_1 + U_2 + U_3$$
The first and second moments of U_1 are:
$$E(U_1) = 0 \cdot (1 - p - 2p) + 1 \cdot 2p + 2 \cdot p = 4p = 0.4$$
and:
$$E(U_1^2) = 0^2 \cdot (1 - p - 2p) + 1^2 \cdot 2p + 2^2 \cdot p = 6p = 0.6$$
Hence, the variance is:
$$Var(U_1) = E(U_1^2) - (E(U_1))^2 = 6p - 16p^2 = 0.6 - 0.16 = 0.44$$
Finally, the mean and variance of U are:
$$E(U) = E(U_1) + E(U_2) + E(U_3) = 12p = 1.2$$
$$Var(U) = Var(U_1) + Var(U_2) + Var(U_3) = 1.32$$

since the number of jobs each user offers is independent of one another.
The value of p that achieves 100% utilization of the four processors required to solve:

$$E(U) = 12p = 4 \quad \Rightarrow \quad p = \frac{1}{3}$$

In general, we have more jobs than servers in the following set of cases for (U_1, U_2, U_3): $(2,2,1)$, $(2,1,2)$, $(1,2,2)$ and $(2,2,2)$, that is, the cases when one user offers only one job but the other two offer two jobs, or when all three users offer two jobs each. The first case occurs with probability:

$$pp(2p) + p(2p)p + (2p)pp = 6p^3 = \frac{6}{27}$$

and case $(2,2,2)$ happens with probability:

$$ppp = p^3 = \frac{1}{27}$$

In total, the probability that more jobs are offered than processors:

$$\frac{7}{27} \quad \text{when } p = \frac{1}{3}$$

Problem 6

We are interested in using slotted Aloha with service differentiation to control medium access with different priority levels. The difference with conventional slotted Aloha is that high-priority users may transmit on a slot with a higher probability than low-priority users.
Consider a network with two stations only: One is of high priority and the other one is of low priority, whereby the transmission probability for high-priority stations is double that for low-priority stations. a) Obtain the transmission probability for each type of station that maximizes the probability of successfully

transmitting any data in a time slot. b) Formulate the problem for the case of three stations: one of high and two of low priority.

Solution

Let $S_i, i = 1, 2$ denote the state of the i-th station: Tx, Idle. There is a successful transmission on a timeslot if only one of the two stations transmits during a time slot. This occurs with probability:

$$P_{success} = P(S_1 = Idle, S_2 = Tx) + P(S_1 = Tx, S_2 = Idle)$$
$$= (1 - 2p)p + 2p(1 - p) = 3p - 4p^2$$

To maximize this quantity, we need to differentiate for p and set that value to zero:

$$\frac{dP_{success}}{dp} = 0 = 3 - 8p \quad \Rightarrow \quad p = \frac{3}{8}$$

We observe that this is a maximum since differentiating again gives a negative value for $p = 3/8$.

When $p = 3/8$, a successful time slot occurs with probability:

$$P_{success} = 3\frac{3}{8} - 4\frac{3^2}{8^2} = 0.5625$$

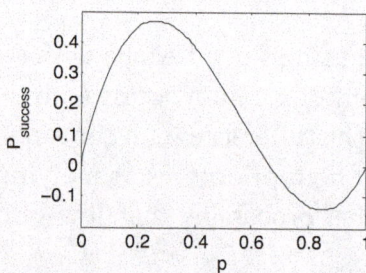

In the case of two low-priority and one high-priority stations, we have:

$$P_{success} = (1 - 2p)\binom{2}{1}p^1(1-p)^{2-1} + 2p\binom{2}{0}p^0(1-p)^{2-0}$$
$$= 4p - 10p^2 + 6p^3$$

again, maximizing this quantity requires:

$$\frac{dP_{success}}{dp} = 0 = 4 - 20p + 18p^2 \quad \Rightarrow \quad p = \{0.2616, 0.8495\}$$

Only the first value maximizes $P_{success} = 0.4695$ since the second value gives a minimum, rather than a maximum.

Problem 7

A group of 2N stations are connected through a slotted Aloha system, where the capture effect is used to solve collisions. This means that, if several stations transmit at the same time, only the signal from the station with the highest power level is valid at the receiver, the other ones are not.

Now, consider a network with 2N stations, divided into N power classes with two stations per class. Assume that each station transmits on a time-slot with probability p. Obtain a) The probability of receiving data from any of the two stations of the highest power level. b) The probability of receiving data from any of the two stations of the lowest power level. c) The probability of a successful transmission over a timeslot, when $N \to \infty$.

Solution

In the first case, we receive data from any of the two highest-power stations only if one of them (but not both) transmits a packet:

$$\binom{2}{1} p^1 (1-p)^{2-1} = 2p(1-p)$$

To receive data from any of the lowest-power stations, it is necessary that none of the other $2N - 2$ stations transmit any packet, and only one of the two lowest-power stations transmits a packet.

This occurs with probability:

$$\binom{2N-2}{0} p^0 (1-p)^{2N-2} \times \binom{2}{1} p^1 (1-p)^{2-1} = 2p(1-p)^{2N-1}$$

There is a collision if two stations of the same power group decide to transmit in the same time slot, while the others from the upper power group do not transmit any packet. So, there is a successful transmission if one station of the first group decides to transmit a packet, or if one station of the second group decides to transmit (while the two stations of the first group do not), and so on. Hence:

$$\begin{aligned} P_{success} &= 2p(1-p) + (1-p)^2 2p(1-p) \\ &\quad + (1-p)^4 2p(1-p) + \ldots \\ &= 2p(1-p) \sum_{n=0}^{\infty} \left((1-p)^2\right)^n = \frac{2(1-p)}{2-p} \end{aligned}$$

Problem 8

In a version of Slotted Aloha, time is divided into long time slots (2 seconds) and short time slots (duration of 1 second), alternating from one another:
Long - Short - Long - Short - Long - Short, etc
Stations can use only the long or the short time slots, but not both. At the beginning of each long slot, a long-slot user transmits with a probability of 1/10 (or remains idle with a probability of 9/10). However, at the beginning of a short time-slot, a short-slot user transmits with probability 1/5. It is observed that five stations use long slots while only three stations use short slots. In addition, those users using long slots use a modulation format that operates at 10 Mb/s when they have a successful transmission, while short-slot users use a more efficient modulation reaching 15 Mb/s.

Obtain (a) the percentage of idle (0 active users), successful (only 1 active user), and collision (2 or more active users) time slots for both long-users and short-slot users; (b) find the average bitrate experienced by long- and short-slot users.

Solution

Concerning long-time slots, we have 5 users transmitting with a probability of $p = 1/10$. Thus:

$$p_{idle}^{(long)} = \binom{5}{0} p^0 (1-p)^5 = \left(\frac{9}{10}\right)^5 = 0.5905$$

$$p_{success}^{(long)} = \binom{5}{1} p^1 (1-p)^4 = 5 \frac{1}{10} \left(\frac{9}{10}\right)^4 = 0.3281$$

$$p_{collision}^{(long)} = 1 - 0.5981 - 0.3281 = 0.0815$$

On the other hand, only three users compete for the short time slots, each attempts to transmit with a probability of $p = 1/5$. Thus:

$$p_{idle}^{(short)} = \binom{3}{0} p^0 (1-p)^3 = \left(\frac{4}{5}\right)^3 = 0.5120$$

$$p_{success}^{(short)} = \binom{3}{1} p^1 (1-p)^2 = 3 \frac{1}{5} \left(\frac{4}{5}\right)^2 = 0.3840$$

$$p_{collision}^{(short)} = 1 - 0.5120 - 0.3840 = 0.1040$$

As observed, users in the short time slots have a higher chance of obtaining a successful transmission (38.4%) than users in the long time slots (32.81%). However, long-time slots have a longer duration (2 seconds) than short-time slots. Thus, in a full Long-Short period of 3 seconds, the average useful time (successful transmissions) is:

$$2 \times 0.3281 + 1 \times 0.3840 = 1.04 \text{ seconds}$$

out of 3 seconds (nearly 1/3 of the total). The remaining 1.96 seconds are wasted in idle periods or collisions.

Given that the five users in the long duration behave the same way, in the long run, each user experiences an equal share (one-fifth) of the total useful time: $2 \times 0.3281 = 0.6562$ seconds divided by 5 users: 0.13124 seconds per user. The same calculus for short-slot users gives 0.3840 seconds equally shared among three users, i.e. 0.128 seconds per user.

Finally, the effective average bandwidth for each type of user follows:

$$B_{\text{eff}}^{(long)} = \frac{0.13124 \text{ s}}{3 \text{ s}} \times 10 \text{ Mb/s} = 0.437 \text{ Mb/s per long-slot user}$$

$$B_{\text{eff}}^{(short)} = \frac{0.128 \text{ s}}{3 \text{ s}} \times 15 \text{ Mb/s} = 0.64 \text{ Mb/s per short-slot user}$$

Short-slot users experience better performance than long-slot users.

Problem 9

An Ethernet switch has 10 computers and 5 servers attached to it, each of them transmits packets to the Internet. Computers transmit 64-byte frames (TCP ACKs) with a probability of 0.6 and 1518-byte frames (Data frames) with a probability of 0.4, while the servers transmit 64-byte frames with a probability of 0.1 and 1518-byte frames with a probability of 0.9. Find the probability that a given selected frame is of size 64 bytes. What is the probability that a server has generated this frame?

Solution

Let random variable FS denote the size of a given frame, with the following possible values $S_{FS} = \{64, 1518\}$ bytes. Now, let E refer to the random variable that denotes the source of a given frame, either Computer or Server. Then, the probability of observing a

64-byte frame equals (total probability theorem):

$$\begin{aligned} P(FS = 64) &= P(FS = 64|E = C)P(E = C) \\ &\quad + P(FS = 64|E = S)P(E = S) \\ &= 0.6 \times \frac{10}{10+5} + 0.1 \times \frac{5}{10+5} = \frac{13}{30} = 0.43 \end{aligned}$$

Then, 43.3% of the frames in the network are short, while the other 56.7% are long 1518-byte frames.

Now, a given short packet comes from a server with probability (Bayes' theorem):

$$\begin{aligned} P(E = S|FS = 64) &= \frac{P(FS = 64, E = S)}{P(FS = 64)} \\ &= \frac{P(FS = 64|E = S)P(E = S)}{P(FS = 64)} \\ &= \frac{0.1 \times \frac{5}{15}}{\frac{13}{30}} = \frac{1}{13} \end{aligned}$$

Problem 10

A WiFi station has two modulation formats to transmit its packets: Mod1 at 1 Mbit/s and Mod2 at 11 Mbit/s. When using Mod2, the frame error probability experienced in the channel is 20%. However, Mod1 is slower but more reliable: the frame error probability is only 3%. An external observer measures that about 12.35% of the frames have been lost. Compute: (a) The percentage of frames transmitted using Mod1 and Mod2; and (2) the percentage of successfully received frames that were transmitted at Mod2.

Solution

First, we know that the two modulation formats have different error probabilities, i.e.: $P(Err|Mod1) = 0.03$ and $P(Err|Mod2) = 0.2$.

Thus, from the total probability theorem, we know that:

$$P(Error) = P(Err|Mod1)P(Mod1) + P(Err|Mod2)P(Mod2)$$
$$= 0.03 \cdot P(Mod1) + 0.2 \cdot P(Mod2) = 0.1235$$

However, $P(Mod1) = 1 - P(Mod2)$ since the station uses one modulation or the other. Then:

$$P(Err) = 0.03 \cdot P(Mod1) + 0.2 \cdot (1 - P(Mod1)) = 0.1235$$

Therefore:

$$P(Mod1) = \frac{0.2 - 0.1235}{0.2 - 0.03} = 0.45$$

and $P(Mod2) = 0.55$.

Thanks to the Bayes' theorem, we obtain the percentage of successful packets that were transmitted using Mod2:

$$P(Mod2|NoErr) = \frac{P(NoErr|Mod2)P(Mod2)}{P(NoErr)}$$
$$= \frac{0.8 \cdot 0.55}{0.8 \cdot 0.55 + 0.97 \cdot 0.45} = 0.502$$

Problem 11

Traffic arrives at a server through two different routes, E and U. On route E, the frames experience a delay that is distributed according to an exponential random variable with a mean of 30 ms. On route U, the traffic suffers a delay that follows a uniformly distributed r.v. between 0 and 100 ms. 40% of the frames are received through Route E and 60% through Route U. (a) Find the average delay of the traffic. (b) Find the probability that the delay is less than 20 ms. (c) If the delay of a frame is less than 20 ms, what is the probability that it arrived via route E? (d) Compute the probability that a randomly chosen frame from route E has a delay smaller than a randomly chosen frame from route U.

Solution

Let X be the delay random variable, let X_e be an exponential random variable with a mean of 30 ms, and let X_u be a random variable that is uniformly distributed between 0 and 100 ms. The average delay is:

$$\begin{aligned} E[X] &= E[X \mid e]P(e) + E[X \mid u]P(u) \\ &= E[X_e]P(e) + E[X_u]P(u) \\ &= 30P(e) + 50P(u) = 42 \text{ ms} \end{aligned}$$

where $P(e) = 0.4$ and $P(u) = 0.6$.
Next, we need to find $P(X < 20)$:

$$\begin{aligned} P(X < 20) &= P(X < 20 \mid e)P(e) + P(X < 20 \mid u)P(u) \\ &= P(X_e < 20)P(e) + P(X_u < 20)P(u) \\ &= (1 - e^{-20/30})P(e) + (20/100)P(u) \\ &= 0.486 \times 0.4 + 0.2 \times 0.6 = 0.315 \end{aligned}$$

Next, we need to use Bayes' theorem:

$$P(e \mid X < 20) = \frac{P(e, X < 20)}{P(X < 20)} = \frac{P(X < 20 \mid e)P(e)}{P(X < 20)}$$

$$= \frac{0.486 \times 0.4}{0.315} = \frac{0.194}{0.315} = 0.619$$

Finally, we can find $\Pr(X_e < X_u)$ using conditioning:

$$P(X_e < X_u) = 1 - P(X_e > X_u)$$

$$\begin{aligned} P(X_e > X_u) &= \int_0^{100} P(X_e > X_u \mid X_u = \tau) f_{X_u}(\tau) d\tau \\ &= \int_0^{100} e^{-\lambda\tau} f_{X_u}(\tau) d\tau = \int_0^{100} e^{-\lambda\tau} \frac{1}{100} d\tau \\ &= \frac{1}{100} \frac{1}{\lambda} e^{-\lambda\tau} \Big|_{100}^{0} = \frac{30}{100} \left(1 - e^{-100/30}\right) = 0.289 \end{aligned}$$

Therefore, $P(X_e < X_u) = 1 - P(X_e > X_u) = 0.711$.

2
The exponential random variable

2.1 Definition

Let X be an exponentially-distributed random variable, characterized by some parameter $\lambda > 0$. The Probability Density Function (PDF) of X follows:

$$f_X(t) = \lambda e^{-\lambda t}, \quad t \geq 0 \qquad (2.1)$$

and has the appearance of the figure below. As shown, the larger the value of λ, the faster the tail decays.

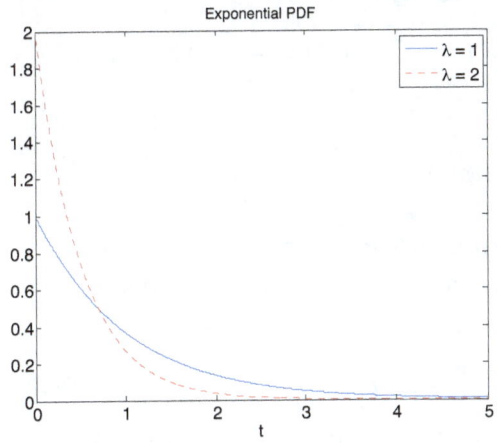

Figure 2.1: Exponential PDF.

The Cumulative Distribution Function (CDF) of X is given by:

$$F_X(t) = P(X \leq t) = \int_{\tau=0}^{t} \lambda e^{-\lambda \tau} d\tau = 1 - e^{-\lambda t}, \quad t \geq 0 \qquad (2.2)$$

and the Complementary Cumulative Distribution Function (CCDF) or Survival function is then:

$$S_X(t) = P(X > t) = 1 - F_X(t) = e^{-\lambda t}, \quad t \geq 0 \qquad (2.3)$$

The mean and variance follow:

$$E(X) = \int_0^\infty t f_X(t) dt = \frac{1}{\lambda} \qquad (2.4)$$

$$Var(X) = \int_0^\infty (t - E(X))^2 f_X(t) dt = \frac{1}{\lambda^2} \qquad (2.5)$$

> **Example 1**
>
> Consider a light bulb whose duration is exponentially distributed with parameter $\lambda = 2$ failures/year. Obtain the average duration of this light bulb and its standard deviation.

> **Solution**
>
> The average duration of this light bulb is:
>
> $$E(X) = \frac{1}{\lambda} = 0.5 \text{ year}$$
>
> and its standard deviation is:
>
> $$Std(X) = \sqrt{Var(X)} = \frac{1}{\lambda} = 0.5 \text{ year}$$

2.2 Simulation and estimation

Consider we have an exponentially distributed random variable X for some parameter λ. Then, we may obtain several observations drawn from X as:

$$o_i = \frac{-\ln u_i}{\lambda}, \quad i = 1, 2, \ldots \qquad (2.6)$$

where the u_i values are uniformly distributed random numbers within the interval $(0,1)$, i.e. $u_i \sim U(0,1)$. The above equation transforms uniformly-distributed random numbers into exponentially-distributed random values with parameter λ.

The figure below shows 40 observations from an exponential random variable with parameter $\lambda = 2$. As shown, most values lie within the range $t \in [0, 1]$. This interval is exactly the mean plus/minus the standard deviation, i.e. $E(X) \pm Std(X)$.

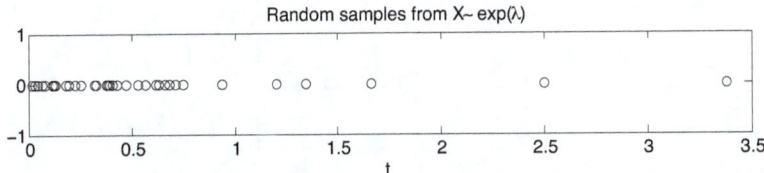

Figure 2.2: Random samples drawn from $exp(\lambda = 2)$

Now, consider a set of observations $O = \{o_1, \ldots, o_N\}$ drawn from an exponential random variable X whose parameter is unknown. The easiest way to estimate its parameter is via the first-moment estimation. Essentially, we know that the expectation of X equals $E(X) = \lambda^{-1}$. Thus, we may obtain the average value of the observations as:

$$\bar{X} = \frac{1}{N} \sum_{i=1}^{N} o_i \qquad (2.7)$$

and approximate parameter λ as:

$$\hat{\lambda} = \frac{1}{\bar{X}} = \left(\frac{1}{N} \sum_{i=1}^{N} o_i \right)^{-1} \qquad (2.8)$$

Now, consider that we are asked to determine whether or not some set of observations $O = \{o_1, \ldots, o_N\}$ were drawn from some exponentially distributed random variable X. The best way to proceed here is to estimate the CCDF (or survival function) of X based on such observations and see whether or not its logarithm follows a straight line. This is based on the fact that the survival function $S_X(t)$ of an exponential random variable follows eq. 2.3, hence its logarithm:

$$\ln S_X(t) = -\lambda t \qquad (2.9)$$

follows a straight line with slope $-\lambda$.

The figure below shows an example of $\ln S_X(t)$ for a set of observations that follow the exponential distribution with parameter $\lambda = 2$.

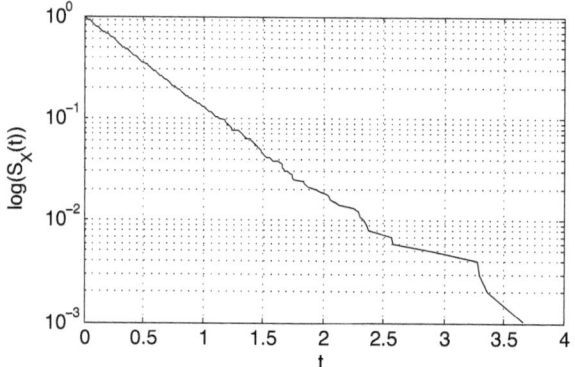

Figure 2.3: Log-survival $log(S_X(t))$ for an exponential PDF

2.3 The memoryless property

The exponential is the only distribution that satisfies the memoryless property which, in mathematical words, states that:

$$P(X > t+s | X > s) = P(X > t), \quad s, t > 0 \qquad (2.10)$$

To prove this property, it is important to observe that:

$$P(X > t+s | X > s) = \frac{P(X > t+s, X > s)}{P(X > s)}$$
$$= \frac{P(X > t+s)}{P(X > s)} = \frac{e^{-\lambda(t+s)}}{e^{-\lambda s}}$$

which turns out to be the same as:

$$P(X > t) = e^{-\lambda t}$$

The memoryless property is only satisfied by the exponential distribution; all other distributions have memory. That is, for all other distributions: $P(X > t+s | X > s) \neq P(X > t)$.

The implications of this property are key to understanding Poisson processes and Continuous-Time Markov Chains, which will be explored in subsequent chapters.

Looking deeper into eq. 2.10, we observe that the memoryless property states that the past s units of time are not important in predicting the future behavior of our exponential random variable X.

Essentially, consider the example of the duration of a light bulb characterized by an exponential distribution with parameter $\lambda = 5$ failures/year. Then, we know that the average duration of this light bulb is $E(X) = \lambda^{-1} = 0.2$ year (or 73 days). Now, consider that we switch on the light on day 0, and we observe that the light bulb works perfectly for $s = 20$ days. If we are asked about the expected duration of this light bulb from day 20 onwards, the answer should be:

My light bulb should last for *another 73 days on average!!!*

This conclusion seems counter-intuitive. One would expect that since the light bulb has already "consumed" 20 days, the remaining life expectancy should be $73 - 20 = 53$ days. This might be true for probability distributions with memory but is wrong for the exponential distribution due to the memoryless property. The next example proves this result. Thanks to the memoryless property, information about the past of an exponential random variable X is not relevant to its future behavior.

Example 2

Consider a light bulb whose duration is characterized by an exponential distribution with parameter $\lambda = 5$ failures/year. Consider that the light bulb has already survived $T = 20$ days. Obtain its average duration assuming $X > T$, i.e. $E(X|X > T)$.

> **Solution**
>
> $E(X|X>T)$ can be obtained as follows:
>
> $$E(X|X>T) = \int_T^\infty \tau f_X(\tau|X>T)d\tau$$
>
> This requires to obtain the conditional PDF first:
>
> $$f_X(X=\tau|X>T) = \frac{P(X=\tau, X>T)}{P(X>T)} = \frac{\lambda e^{-\lambda\tau}}{e^{-\lambda T}} = \lambda e^{-\lambda(\tau-T)}, \quad \tau \geq T$$
>
> Hence, the integral above becomes:
>
> $$E(X|X>T) = \int_T^\infty \tau \lambda e^{-\lambda(\tau-T)} d\tau$$
>
> which, after some calculus, yields:
>
> $$E(X|X>T) = T + \frac{1}{\lambda} \text{ days}$$
>
> Hence, the expected total duration of the light bulb is $T + E(X) = 20 + 73 = 93$ days, that is, another $E(X) = 73$ days more than the already consumed $T = 20$ days.

In conclusion, the memoryless property states that the previously observed duration of the light bulb is not important in determining its future behavior. The remaining lifetime of the light bulb is again exponentially distributed with parameter λ.

2.4 Minimum of several exponential random variables

Another important property of the exponential distribution concerns the minimum of several exponential random variables:

Minimum Let $X_1, \ldots, X_N$ be independent exponential random variables characterized by parameters $\lambda_1, \ldots, \lambda_N$ respectively. Then, the minimum of them, denoted as:

$$X_m = \min\{X_1, \ldots, X_N\}$$

is also exponentially distributed with parameter:

$$\lambda_m = \lambda_1 + \ldots + \lambda_N$$

Essentially:

$$P(X_m > t) = P(X_1 > t, X_2 > t, \ldots, X_N > t) \quad (2.11)$$

which implies that all random variables X_i have to be greater than t to satisfy that the minimum of them is also greater than t. Both requirements are equivalent. Then:

$$P(X_m > t) = P(X_1 > t)P(X_2 > t) \cdots P(X_N > t) = e^{-\sum_{i=1}^{N} \lambda_i t}$$
$$(2.12)$$

which proves that the random variable X_m is also exponentially distributed with parameter $\lambda_m = \sum_{i=1}^{N} \lambda_i$.

> ### Example 3
>
> Consider a car with three main components subject to failure: its engine, its electricity system, and its air conditioning, each characterized by an exponential random variable with parameters: $\lambda_e = 1$ failure per decade, $\lambda_{es} = 2$ failures per decade and $\lambda_{ac} = 5$ failures per decade respectively. Find the probability of having one failure of any type in its first year.
>
> ### Solution
>
> Let X_m denote the first failure of the car. This first failure is the minimum of the three exponential random variables that characterize the car's engine, electricity system, and air conditioning. Thus, X_m is exponentially distributed with parameter λ_m:
> $\lambda_m = \lambda_e + \lambda_{es} + \lambda_{ac} = 1 + 2 + 5 = 8$ failure/decade, or 0.8 failure/year. Hence, we have the first failure in less than a year with probability:
>
> $$P(X_m \leq 1 \text{ year}) = 1 - e^{-0.8 \times 1} = 0.55$$
>
> It is finally worth remarking that the time units must match with parameter λ. If instead of using years as time units we use

decades, then the correct calculus would be:

$$P(X_m \leq 0.1 \text{ decade}) = 1 - e^{-8 \times 0.1} = 0.55$$

where we have used parameter $\lambda = 8$ failure/decade in order to match the 0.1 decade time value.

2.5 Comparison of several exponential random variables

Comparison Consider two independent exponential random variables X_1 and X_2 characterised by parameters λ_1 and λ_2. Then:

$$P(X_1 < X_2) = \frac{\lambda_1}{\lambda_1 + \lambda_2} \quad (2.13)$$

This result arises after applying the total probability theorem for all possible values of X_2:

$$\begin{aligned} P(X_1 < X_2) &= \int_0^\infty P(X_1 < X_2 = t | X_2 = t) P(X_2 = t) dt \\ &= \int_0^\infty \left(1 - e^{-\lambda_1 t}\right) \lambda_2 e^{-\lambda_2 t} dt = \frac{\lambda_1}{\lambda_1 + \lambda_2} \end{aligned}$$

This property allows us to compare exponential random variables very easily by using the ratio of their rates.

> ### Example 4
>
> Consider the same car as in the previous example. Find the probability that the engine fails before the other two components in the car. Find also the probability that the engine fails twice in a row, that means, the first and second failure also occurs in the engine (the engine is assumed to be repaired after the first failure).

Solution

The engine fails before the other two components with probability:

$$P(X_e \leq \min\{X_{es}, X_{ac}\}) = \frac{\lambda_e}{\lambda_e + (\lambda_{es} + \lambda_{ac})} = \frac{1}{1+2+5} = \frac{1}{8}$$

Here, we have also applied the property that a minimum of two exponential random variables are also exponentially distributed. Next, the engine fails twice in a row with probability: $\frac{1}{8} \times \frac{1}{8}$. Essentially, once the first failure occurs, the system renews itself and starts all over again, thanks to the memoryless property of the exponential distribution. In other words, after the first engine failure, the next failure does not take into account the history or past of the air conditioning or electricity system (memorylessness).

The properties of minimum and comparison of exponential random variables have further applications in the Poisson process and Continuous-Time Markov Chains.

2.6 Further problems

Problem 1

Prove that the uniform distribution does not have memoryless properties.

Solution

This requires to check whether or not:

$$P(X > t + s | t > s) = P(X > t)$$

holds for the uniform distribution $U(0, a)$, i.e.:

$$f_X(t) = \frac{1}{a}, \quad 0 \leq t \leq a$$

First, the CDF of a uniform distribution $X \sim U(0, a)$ follows:

$$F_X(t) = P(X \leq t) = \int_0^t f_X(\tau)d\tau = \frac{t}{a}, \quad 0 \leq t \leq a$$

and its CCDF:
$$S_X(t) = P(X > t) = 1 - \frac{t}{a}, \quad 0 \leq t \leq a$$

Thus:
$$P(X > t+s | X > s) = \frac{P(X > t+s, X > s)}{P(X > s)} = \frac{P(X > t+s)}{P(X > s)} = \frac{1 - \frac{t+s}{a}}{1 - \frac{s}{a}}$$

which is very different from:
$$S_X(t) = P(X > t) = 1 - \frac{t}{a}$$

Problem 2

Consider a random variable X_1 uniformly distributed in the range $(0, a)$ and a second random variable X_2 exponentially distributed with parameter λ. Compute $P(X_1 < X_2)$

Solution

This requires solving:
$$P(X_1 < X_2) = \int_{t=0}^{\infty} P(X_1 < t | X_2 = t) P(X_2 = t) dt$$
$$= \int_{t=0}^{a} \frac{t}{a} \lambda e^{-\lambda t} dt + \int_{t=a}^{\infty} 1 \cdot \lambda e^{-\lambda t} dt$$

Solving them separately brings:
$$P(X_1 < X_2) = \frac{1}{\lambda a} \left(1 - e^{-\lambda a}\right)$$

Problem 3

Consider a network switch with 24 input ports. Each port suffers failures following an exponential distribution with parameter $\lambda = 4$ failure/day. (a) Obtain the average time elapsed between one failure and the next. (b) Obtain the probability of having at least 1 failure in one hour.

Next, consider that the first 10 input ports are connected to lab computers, whereas the other 14 are connected to other networks; (c) obtain the probability that the next four consecutive failures affect the connectivity of lab computers. (d) Consider that 10 failures have occurred in a day and obtain the probability that six of them affect lab computers.

Solution

After a failure occurs, we know that the next failure event is characterized as the minimum of 24 exponential random variables, thus also exponentially distributed with parameter $\lambda_m = \sum_i \lambda_i = 24 \times 4 = 96$ failure/day or 4 failure/hour. Hence, the average time elapsed until the next failure is:

$$E(X_m) = \frac{1}{\lambda_m} = \frac{1}{4} \text{ hour (or 15 mins)}$$

The probability of one or more failures in one hour can be obtained as the complementary of having no failures in one hour. Consider we take an initial time instant t and we observe the number of failures until $t + 1$ hour. Since the past is not important, the first failure after t is exponentially distributed with parameter 24λ. We have no failures within $(t, t + 1)$ hour if the first failure occurs any time after $t + 1$, i.e.:

$$P(X_m > 1) = e^{-4 \times 1} = 0.0183$$

Thus, the complementary is:

$$1 - P(X_m > 1) = 0.9817$$

In the second scenario, we have that lab failures are then characterized by an exponential distribution with parameter $10\lambda = 40$ failure/day, whereas the other failures are also exponentially distributed with parameter $14\lambda = 56$ failure/day. The probability that any port attached to a lab computer fails before any port attached to other networks follows:

$$P(X_{lab} < X_{other}) = \frac{40}{40 + 56} = \frac{5}{12}$$

Since the system renews itself at all times (thanks to the memoryless property of the exponential distribution), the second failure comes from a lab-computer port with probability 5/12 again, and so on. Thus, we have four consecutive lab failures with probability:

$$\left(\frac{40}{40+56}\right)^4 = 0.03$$

Finally, under the assumption that 10 failures have occurred during some period, and each of them affects a lab computer with probability $p = \frac{40}{40+56} = \frac{5}{12} = 0.42$, then we have that exactly six of them affect lab computers with probability:

$$\binom{10}{6} p^6 (1-p)^{10-6} = 0.13$$

Problem 4

Consider a network switch with 24 input ports, each one receives packets following a Poisson process with rate λ. We say that two packets belong to a burst if their inter-arrival time is smaller than some amount of time T. Find the probability that the switch receives a burst of exactly 10 packets from one of its input ports. Repeat the exercise for a burst of n packet arrivals at the switch from any of its input ports.

Solution

We have a burst of exactly 10 packets if the following condition is met:

$$(X_2 < T) \cap (X_3 < T) \cap \ldots \cap (X_{10} < T) \cap (X_{11} > T)$$

which means that, after the first packet's arrival, the nine subsequent packets meet the requirement of inter-arrival time smaller than T, while the eleventh packet does not.

Since the packet arrivals are independent, then:

$$P(B = 10) = (P(X_2 < T))^9 P(X_{11} > T) = \left(1 - e^{-\lambda T}\right)^9 e^{-\lambda T}$$

where B denotes the random variable that accounts for the size of the burst.

In the second case, the aggregated process of all 24 input ports is also Poisson with rate 24λ. Then,

$$P(B = n) = \left(1 - e^{-24\lambda T}\right)^{n-1} e^{-24\lambda T}$$

as it follows from the Binomial distribution.

Problem 5

Let us consider a hacker who can break passwords in an exponentially distributed random time X with a 100-day mean. Consider that user Jimmy changes his password on the first day of January. (a) Find the probability that the hacker will break his password in a whole year. (b) Repeat the exercise if user Jimmy changes his password on the first day of every month.

Solution

This hacker can break passwords in an exponentially distributed random time with parameter $\lambda = \frac{1}{E(X)} = \frac{1}{100} = 0.01$ password/-day. Thus, the hacker will find the user's password before the end of the year with probability:

$$P(X < 365) = 1 - e^{-\frac{1}{100} \times 365} = 1 - e^{-3.65} = 0.974$$

In the second case, the probability that the hacker finds the user's password in a month follows:

$$P\left(X < \frac{365}{12}\right) = 1 - e^{-\frac{1}{100}\frac{365}{12}} = 1 - e^{-\frac{365}{1200}} = 0.2623$$

Changing the password every month makes it more difficult for the hacker to break each password. However, the hacker has 12 chances in a year (one password per month) to find this new password. So, the hacker will find any of the 12 passwords with the following probability:

$$\sum_{i=1}^{12} \binom{12}{i} p^i (1-p)^{12-i} = 1 - \binom{12}{0} p^0 (1-p)^{12-0} = 1 - (1-p)^{12}$$

where the value of p equals the probability of finding a given password in less than one month obtained above:

$$p = P\left(X < \frac{365}{12}\right) = 1 - e^{-\frac{365}{1200}} = 0.2623$$

which yields:

$$1 - \left(1 - \left(1 - e^{-\frac{365}{1200}}\right)\right)^{12} = 1 - e^{-3.65} = 0.974$$

which is the same result as case one.
Again, this is another consequence of the memoryless property of the exponential distribution. After one month, the probability of breaking a password does not depend on the past. It does not matter if it is a new password or an old one, the hacker sees the same probability of breaking a password after one month.

Problem 6

Repeat the previous problem under the assumption that the hacker requires a uniformly distributed random time with the same mean: 100 days.

Solution

First of all, it is worth remarking that the CDF of a $U(0, a)$ distribution follows:

$$F_X(t) = P(X \leq t) = \begin{cases} t/a & \text{if } 0 \leq t \leq a, \\ 1 & \text{if } t \geq a. \end{cases}$$

and its mean is:

$$E(X) = \frac{a}{2}$$

Thus, the hacker requires a uniform random time $X \sim U(0, 200)$ (i.e. mean is $100 = \frac{a}{2}$ days) to break a password. The probability of breaking a password in less than one year is then:

$$P(X \leq 365) = 1$$

so the hacker always manages to break the user's password, since the maximum time taken to do so is 200 days.
If the user changes his/her password every month, then the hacker breaks a monthly password with probability:

$$P\left(X < \frac{365}{12}\right) = \frac{365/12}{200} = 0.1521$$

Because the hacker has 12 chances to break a monthly password, then the probability of breaking at least one follows:

$$\sum_{i=1}^{12} \binom{12}{i} p^i (1-p)^{12-i} = 1 - \binom{12}{0} p^0 (1-p)^{12-0}$$
$$= 1 - (1-p)^{12}$$

where $p = P\left(X < \frac{365}{12}\right) = 0.1521$. Thus:

$$1 - (1 - 0.1521))^{12} = 0.8619$$

Problem 7

Consider a DNS service with three servers attending requests. The time required by each server to attend a request can be modeled with an exponential distribution X_i, $i = 1, 2, 3$, with different mean values: 0.5, 1, and 2 ms respectively. Consider that the three servers are busy upon the arrival of a new request. (a) Find the probability that the new request has to wait more than 2 ms before it can be attended; (b) find the average total time spent by this new request in the DNS service.

Solution

Thanks to the memoryless property of the exponential distribution, we know that the new request observes three exponentially

distributed random variables upon its arrival, no matter how long the three servers have been serving their current request.
In other words, the previous time spent by the three servers is not important to the new request's point of view, this one sees three exponential random variables X_1, X_2, and X_3. Furthermore, the minimum between these three random variables is also exponentially distributed with rate:

$$\lambda_{min} = \lambda_1 + \lambda_2 + \lambda_3 = \frac{1}{0.5} + \frac{1}{1} + \frac{1}{2} = 3.5 \text{ request/ms}$$

Thus, the probability that the new request has to wait more than 2 ms before being attended is:

$$P(X_{min} > 2ms) = e^{-\lambda_{min} 2} = e^{-3.5 \cdot 2} = e^{-7}$$

Note that the time units of λ and t must match.

Secondly, the average time spent by the new request in the DNS service is the sum of two components: $E(W) + E(S)$ which refers to the average waiting time until one server is free and the average time spent by the request in that server. The first amount is:

$$E(W) = \frac{1}{\lambda_{min}} = \frac{1}{3.5} = \frac{2}{7} \text{ ms}$$

The second amount arises after conditioning on that particular server that attends this new request:

$$E(S) = \sum_{i=1}^{3} E(S|Server = i) P(Server = i)$$

$$= E(X_1) \frac{\lambda_1}{\lambda_1 + \lambda_2 + \lambda_3} + E(X_2) \frac{\lambda_2}{\lambda_1 + \lambda_2 + \lambda_3}$$

$$+ E(X_3) \frac{\lambda_3}{\lambda_1 + \lambda_2 + \lambda_3}$$

$$= 0.5 \frac{\frac{1}{0.5}}{\frac{1}{0.5} + \frac{1}{1} + \frac{1}{2}} + 1 \frac{\frac{1}{1}}{\frac{1}{0.5} + \frac{1}{1} + \frac{1}{2}} + 2 \frac{\frac{1}{2}}{\frac{1}{0.5} + \frac{1}{1} + \frac{1}{2}}$$

$$= \frac{6}{7} \text{ ms}$$

Hence, the total time spent until the request is attended and served follows:

$$E(W) + E(S) = \frac{2}{7} + \frac{6}{7} = \frac{8}{7} \text{ ms}$$

Note that this problem is somehow related to queueing theory, which will be further studied in subsequent chapters.

Problem 8

In a P2P backup system, files are divided into chunks and replicated over different machines. For instance, a file of 1 MByte in size is divided into 1000 chunks of 1 KByte/chunk. A number $R = 10$ of copies of every chunk are stored on different computers on the Internet. We assume that each computer removes chunks at a rate 0.05 chunks/day. Find the probability that a 1 MByte file survives more than 10 days.

Solution

Let $F_1, \ldots, F_N$ refer to the N chunks of the file, and consider C_{ij} refers to the i-th copy of the j-th chunk for this file. Hence, we have $C_{11}, \ldots, C_{R1}$ copies of the first chunk, $C_{12}, \ldots, C_{R2}$ copies of the second chunk, etc.

The file survives if at least one copy of each chunk survives, i.e.:

$$P(\text{File survives}) = P(F_1 \text{ survives}, F_2 \text{ survives}, \ldots, F_N \text{ survives})$$
$$= (P(F_1 \text{ survives}))^N$$

Now, F_1 survives if at least one of its copies survives. In other words:

$$P(F_1 \text{ survives}) = 1 - \binom{R}{0} p^0 (1-p)^{R-0} = 1 - (1-p)^R$$

where p refers to the probability that a copy of a chunk survives beyond T. This is:

$$p = P(C_{11} \text{ survives beyond } T) = P(C_{11} > T) = e^{-\lambda T}$$

Hence, the whole file survives T days with probability:

$$P(\text{File survives}) = \left(1 - \left(1 - e^{-\lambda T}\right)^R\right)^N$$

In our case, $N = 1000$ chunks, $R = 10$ copies per chunk, $\lambda = 0.05$ chunk removal per day and $T = 10$ days:

$$P(\text{File survives}) = \left(1 - \left(1 - e^{-0.05 \cdot 10}\right)^{10}\right)^{1000} = 0.9149$$

3
The Poisson process

3.1 Counting processes

Counting process A counting process $N(t)$, $t \geq 0$ is a random process that counts the number of events that occurred within the time interval $(0, t)$.

The next figure shows a realization example of a counting process. The example shows only the first four seconds of the process, and six events, $e_1, \ldots, e_6$ occurred within such time. Hence, $N(4s) = 6$ events. We can also observe that $N(3s) = 5$ events, $N(2s) = 3$ events, etc.

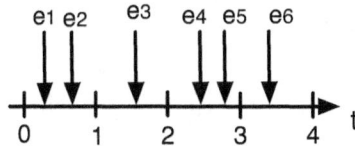

Figure 3.1: Realisation of a counting process

The events e_i may refer to packet arrivals to a network router, node failures in a network, or any other type of event. Every counting process $N(t)$ must satisfy the following properties:

1. $N(t) \geq 0$

2. $N(t)$ is integer-valued

3. If $s < t$, then $N(s) \leq N(t)$

The value $N(t) - N(s)$ refers to the number of events that occurred within the interval (s, t). In the previous example process, we observe:

$$N(4s) - N(3s) = 6 - 5 = 1 \text{ event}$$

3.2 Stationary and independent increments

In a counting process, it is important to determine whether or not the following two properties are met:

Independent increments if the number of events occurred in disjoint time intervals are independent.

Stationary increments if the probability distribution of the number of events that occurred on a time interval only depends on its length.

In the next example, we say that the process has independent increments if the number of events occurred within the time intervals (t_0, t_1) and (t_2, t_3) are probabilistically independent. This must hold for every pair of disjoint time intervals (s, t) and (u, v) for any $s, t, u, v \geq 0$.

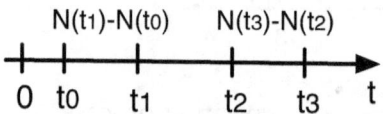

Figure 3.2: Independent and stationary increments. Here: $t_1 - t_0 = t_3 - t_2$

On the other hand, we say that the process has stationary increments if the number of events occurred within the time intervals (t_0, t_1) and (t_2, t_3), which have the same length, have the same probability distribution. This must hold again for any pair of time intervals $(s, s+t)$ and $(u, u+t)$ for any $s, t, u \geq 0$.

3.3 The Poisson process

Poisson process A Poisson process $N(t)$ with rate $\lambda > 0$ is a counting process that satisfies:

1. $N(0) = 0$
2. $N(t)$ has independent and stationary increments
3. The number of events that occurred within a time interval of length t follows the Poisson distribution with rate λt:

$$P(N(t) = n) = \frac{(\lambda t)^n}{n!} e^{-\lambda t}, \quad n = 0, 1, \ldots, \quad t \geq 0 \quad (3.1)$$

The Poisson process has the following mean and variance:

$$E(N(t)) = \lambda t = Var(N(t)) \quad (3.2)$$

Example 1

Consider a Poisson process characterized by parameter $\lambda = 2$ event/s. Obtain the probability that exactly five events occur within the time interval $(2, 4)$ secs. Find the probability of one event or more within $(1, 2)$ secs.

Solution

Thanks to the fact that the Poisson process has stationary increments, the number of events within $(2, 4)$ and $(0, 2)$ secs are probabilistically identical. Hence:

$$P(N(2s) = 5) = \frac{(2 \cdot 2)^5}{5!} e^{-2 \cdot 2} = 0.156$$

The probability of one or more events within $(1, 2)$ secs can be computed as the complementary of the probability of no events in

(1, 2). Again, thanks to the property of stationary increments, we have that:

$$P(N(1s) \geq 1) = 1 - P(N(1s) = 0) = 1 - \frac{(2 \cdot 1)^0}{0!} e^{-2 \cdot 1} = 0.865$$

Example 2

Can you estimate the rate λ of the next process?

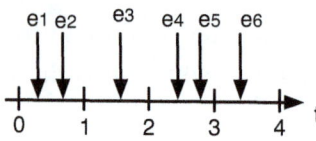

Solution

From the figure, we observe that there are exactly 6 events within $(0, 4)$ secs. Hence we can use this information for estimating the process' rate λ as (eq. 3.2):

$$\hat{\lambda} = \frac{E(N(t))}{t} = \frac{6 \text{ event}}{4 \text{ s}} = 1.5 \text{ event/s}$$

Alternatively, we may select disjoint intervals of one second and obtain the average number of event arrivals on these intervals. For instance, we observe the following number of arrivals in different intervals of one-second length:

$$(0s, 1s) \rightarrow 2 \text{ events}$$
$$(1s, 2s) \rightarrow 1 \text{ event}$$
$$(2s, 3s) \rightarrow 2 \text{ events}$$
$$(3s, 4s) \rightarrow 1 \text{ event}$$

The mean is then $\frac{2+1+2+1}{4} = 1.5$ events per interval length of one second (event/s).

3.4 Inter-arrival times of events

In the next figure, X_1 refers to the time until the first event e_1 occurs, X_2 refers to the inter-arrival time between the first and the second events, and so on. In other words, the random variables X_i, $i = 2, \ldots$ refer to the time elapsed between events e_{i-1} and e_i, and X_1 refers to the time until the first event arrival.

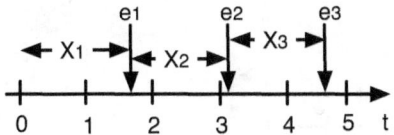

Figure 3.3: Inter-arrival times X_i, $i = 1, 2, \ldots$ of events.

In a Poisson process $N(t)$ with rate λ event per unit of time, the inter-arrival times X_i between events follow an exponential distribution with parameter λ.

This is proved for the first event noting that the following two conditions are equivalent:

$$N(t) = 0 \iff X_1 > t$$

which refers to the fact that no event occurs within $(0, t)$ if and only if the first event occurs after time t. Then:

$$P(X_1 > t) = P(N(t) = 0) = \frac{(\lambda t)^0}{0!} e^{-\lambda t} = e^{-\lambda t}$$

which is the CCDF of an exponential random variable with parameter λ.

The same reasoning applies to the second event: Consider that the the first event occurs at some time s. Then, the second event occurs after time $s + t$ only if no events occur within $(s, s + t)$:

$$\begin{aligned} P(X_2 > s + t | X_1 = s) &= P(\{N(s+t) - N(s)\} = 0) \\ &= P(N(t) = 0) = \frac{(\lambda t)^0}{0!} e^{-\lambda t} \\ &= e^{-\lambda t} \end{aligned}$$

which is again the CCDF of an exponentially distributed random variable with parameter λ. Here we have applied the property of stationary increments of the Poisson process. This reasoning can be further applied to prove $X_3, X_4, \ldots$.

> **Example 3**
>
> Can you estimate the average inter-arrival time of events for the Poisson process of the next figure?
>
>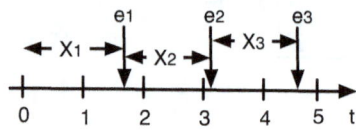

> **Solution**
>
> From the figure, we observe that exactly three events have arrived within $(0, 5s)$. With this information, we can proceed to estimate the rate λ of this process as we did in the previous example:
> $$\hat{\lambda} = \frac{E(N(t))}{t} = \frac{3}{5} = 0.6 \text{ event/s}$$
> Thus, the average inter-arrival time between events is:
> $$E(X_i) = \frac{1}{\lambda} = \frac{5}{3} = 1.67 \text{ secs}$$
> which matches the average inter-arrival times observed in the figure.

3.5 Waiting times

Now, consider the waiting or absolute arrival times S_i of events, as shown in the next figure. Essentially, the values of

S_i follow:

$$S_1 = X_1$$
$$S_2 = X_1 + X_2$$
$$S_3 = X_1 + X_2 + X_3$$
(3.3)

and so on. In general:

$$S_i = \sum_{j=1}^{i} X_j \qquad (3.4)$$

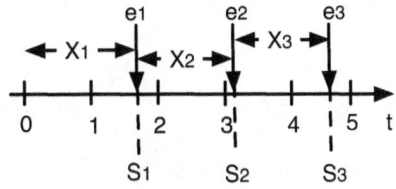

Figure 3.4: Absolute arrival times.

So, the absolute arrival times of events comprise the sum of a number of exponentially distributed random variables with parameter λ. Such a sum is known to follow the Gamma distribution with parameter λ and i degrees of freedom: $S_i \sim \Gamma_t(i, \lambda)$, where:

$$f_{S_i}(t) = \Gamma_t(i, \lambda) = \frac{\lambda^i t^{i-1}}{(i-1)!} e^{-\lambda t}, \quad t \geq 0 \qquad (3.5)$$

with the following mean and variance:

$$E(S_i) = \frac{i}{\lambda}, \quad Var(S_i) = \frac{i}{\lambda^2} \qquad (3.6)$$

It can be easily proved that $\Gamma(1, \lambda)$ is an exponential PDF with parameter λ.

Example 4

Can you find the average arrival time of the third event for the Poisson process of Fig. 3.3? Find also its standard deviation.

Solution

After estimating, the process' rate:

$$\hat{\lambda} = \frac{E(N(t))}{t} = \frac{3}{5} = 0.6 \text{ event/s}$$

we can say that the third arrival is Gamma-distributed:

$$S_i \sim \Gamma(i=3, \lambda = 0.6)$$

The average arrival time of the third event, and its standard deviation, are:

$$E(S_3) = \frac{3}{0.6} = 5s, \quad Std(S_3) = \sqrt{\frac{3}{0.6^2}} = 2.89s$$

So the third event should most likely lie within 5 ± 2.89 seconds.

3.6 Conditional distribution of arrivals

Now, consider that we are told that exactly one event occurred at some point within the time interval $(0, t)$. The question now is: when is it more likely to have occurred: at the beginning or at the end of such a time interval? Let $Y_1 = s, s \in (0, t)$ denote the exact arrival time of such an event. The conditional distribution of Y_1 under the assumption that there is exactly one event within $(0, t)$ is:

$$P(Y_1 < s | N(t) = 1), \quad 0 \leq s \leq t$$

This probability can be obtained by noting that:

$$P(Y_1 < s | N(t) = 1) = \frac{P(Y_1 < s, N(t) = 1)}{P(N(t) = 1)}$$

$$= \frac{P(1 \text{ event in}(0, s)) P(0 \text{ event in}(s, t))}{P(1 \text{ event in}(0, t))}$$

(3.7)

where the joint probability $P(Y_1 < s, N(t) = 1)$ requires to have exactly one event before s and zero events after s. This yields:

$$P(Y_1 < s | N(t) = 1) = \frac{\frac{(\lambda s)^1}{1!} e^{-\lambda s} \times \frac{(\lambda(t-s))^0}{0!} e^{-\lambda(t-s)}}{\frac{(\lambda t)^1}{1!} e^{-\lambda t}}$$

$$= \frac{s}{t}, \quad 0 \le s \le t \quad (3.8)$$

which is the CDF of a uniform distribution $U(0, t)$.

In conclusion, if we are told that an event has occurred over some time interval, then all possible locations for that event are equally likely.

3.7 Aggregation and sampling of Poisson processes

Aggregation Let $N_1(t)$ and $N_2(t)$ be Poisson processes with rates λ_1 and λ_2 respectively. Then, the aggregated process $N_{agg}(t)$ which contains the events of $N_1(t)$ and $N_2(t)$ is also a Poisson process with rate $\lambda_1 + \lambda_2$.

Sampling Let $N(t)$ be a Poisson process with rate λ, and let $0 \le p \le 1$ be some sampling probability. Then, the process obtained by sampling the events of $N(t)$ with probability p is also a Poisson process with rate λp. In addition, the events left apart is also a Poisson process with rate $\lambda(1-p)$.

In the next figure, we have two Poisson processes, $N_1(t)$ and $N_2(t)$, whose rates are λ_1 and λ_2 respectively. The combined (or aggregated) process is also Poisson with rate $\lambda_1 + \lambda_2$.

> ### Example 5
>
> Consider an Ethernet switch with 24 input ports and two output ports. Each input port receives packets following a Poisson process with rate $\lambda_i = 10$ packet/s. The switch is observed to forward about 70% of the

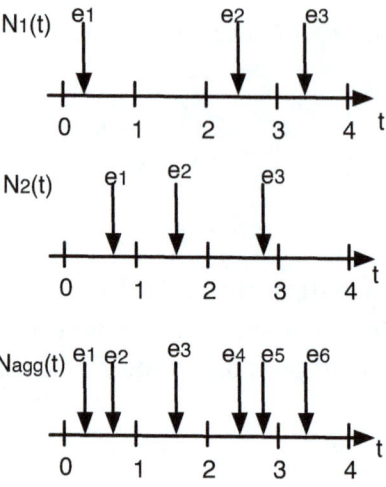

Figure 3.5: Aggregation of two processes

packets to the first output port, while the other 30% are shown to go to the second output port. Find the rates of the Poisson processes in both output ports.

Solution

The two processes at the output ports of the switch comprise the aggregation and sampling of 24 Poisson processes, hence the two processes are also Poisson. Their rates are:

$$\lambda_{o_1} = \sum_{i=1}^{24} \lambda_i p = 240 \times 0.7 = 168 \text{ packet/s}$$

$$\lambda_{o_2} = \sum_{i=1}^{24} \lambda_i (1-p) = 240 \times 0.3 = 72 \text{ packet/s}$$

3.8 The theorem of Palm-Khintchine

Theorem of Palm-Khintchine The aggregation of a large number of not necessarily Poissonian renewal processes, each with a small intensity has Poissonian properties.

The theorem of Palm-Khintchine provides a very useful result for many communication scenarios, where you have the superposition of many independent low-intensity non-Poisson processes, for example, telephony networks.

> ### Example 6
>
> Consider an Ethernet switch with 64 input ports, attached to independent traffic sources, and one output port that forwards all traffic to an upper node. Each traffic source is observed to offer packets following a counting process with uniformly distributed packet inter-arrival times $U(2, 12)$ ms. Obtain the aggregated packet process in the output port.

> ### Solution
>
> In this example, the individual processes are not Poisson since their packet inter-arrival times are not exponentially distributed, but uniformly distributed. The average event inter-arrival time is:
>
> $$E(X_i) = \frac{b+a}{2} = \frac{12+2}{2} = 7 \text{ ms}$$
>
> The average rate of packets for an individual process is then:
>
> $$\frac{1 \text{ packet}}{E(X_i) \text{ s}} = 142.86 \text{ packet/s}$$
>
> The aggregation of a 64 independent counting process whose individual rates are 142.86 packet/s should produce a Poisson process with an aggregated rate:
>
> $$\lambda_{agg} = 64 \times 142.86 = 9142.86 \text{ packet/s}$$
>
> as it follows from the Palm-Khintchine's theorem. A remark from the previous chapter that the logarithm of the survival function of an exponentially distributed random variable X follows a straight line with slope $-\lambda_{agg}$:
>
> $$\begin{aligned} \ln S_X(t) &= \ln\left(e^{-\lambda_{agg} t}\right) \\ &= -\lambda_{agg} t \end{aligned}$$

Indeed, the best way to check this assumption is by taking the inter-arrival times between packets from the aggregated process and observing whether or not they are exponentially distributed with rate λ_{agg}. The next figure shows the log-survival function of the inter-arrival times of packets in the aggregated process. As shown, the process has Poisson properties since the log-survival plot approximates a straight line.

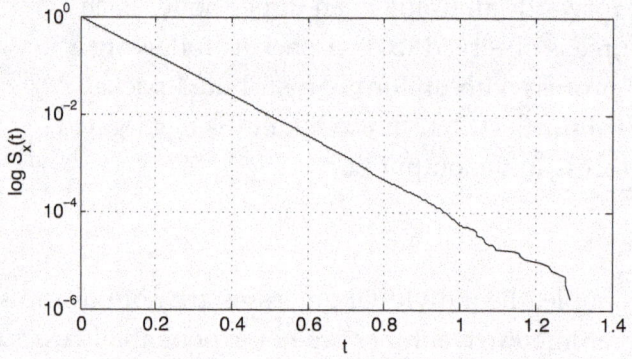

3.9 Further problems

Problem 1

Consider a counting process where event inter-arrival times are uniformly distributed $U(0,2)$ secs. Find the probability values: $P(N(1s) = 0)$ and $P(N(1s) = 1)$, that is, the probability of exactly 0 events within $(0,1)$ second, and the probability of exactly 1 event within the same time interval.

Solution

First of all, it is worth remarking that the PDF of the uniform distribution $U(0,a)$ is $f_X(x) = \frac{1}{a}$, and its CDF is $F_X(x) = \frac{x}{a}$, for $0 \leq x \leq a$.

Essentially:
$$N(t) = 0 \iff X_1 > t$$
In other words, there is no event arrival within $(0, t)$ if the the first event arrives after time t, thus:
$$P(N(1s) = 0) = P(X_1 > 1s) = 1 - \frac{1}{2} = 0.5$$
There is exactly 1 event within $(0, t)$ if the first event arrives before time t and the second event arrives after time t, both conditions need to be met. This means:
$$P(N(1s) = 1) = P(X_1 < 1s, X_1 + X_2 > 1s)$$
To solve this, we first need to obtain the joint distribution of X_1 and X_2:
$$f_{X_1,X_2}(x_1, x_2) = \frac{1}{2} \times \frac{1}{2} = \frac{1}{4}, \quad 0 \leq x_1, x_2 \leq 2s$$
since both random variables X_1 and X_2 are independent. Hence, the cases where $X_1 < 1s$ and $X_1 + X_2 > 1s$ require to solve the next integral:
$$\int_{x_1=0}^{1} \int_{x_2=1-x_1}^{2} f_{X_1,X_2}(x_1, x_2) dx_1 dx_2$$
First, we solve the integral for x_2:
$$\int_{x_1=0}^{1} dx_1 \int_{x_2=1-x_1}^{2} \frac{1}{4} dx_2 = \int_{x_1=0}^{1} \frac{1+x_1}{4}$$
and then for x_1:
$$\int_{x_1=0}^{1} \frac{1+x_1}{4} dx_1 = \frac{3}{8}$$
Thus:
$$P(N(1s) = 1) = \frac{3}{8} = 0.375$$
The probability of two arrivals within $(0, t)$, i.e. $P(N(t) = 2)$ would follow the same procedure: first, find the joint PDF of X_1, X_2, X_3, and secondly solving the integral for the cases where $X_1, X_2 < 1s$ and $X_3 > 1s$.

Problem 2

Consider two Poisson processes $N_1(t)$ and $N_2(t)$ with rates λ_1 and λ_2 respectively. The next figure shows a realization of each process. Estimate the Processes' rates from the information shown in the figure.

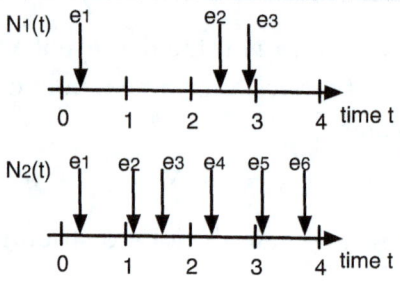

Solution

Clearly, $\lambda_2 > \lambda_1$ since the second process has more events than the first one over the same amount of time: 4 seconds.
Now, their rates can be estimated from:

$$\hat{\lambda}_1 = \frac{E(N(t))}{t} = \frac{E(N(4s))}{4s} = \frac{3}{4} = 0.75 \text{ event/s}$$

$$\hat{\lambda}_2 = \frac{E(N(4s))}{4s} = \frac{6}{4} = 1.5 \text{ event/s}$$

Problem 3

Consider an Ethernet switch that receives packets following a Poisson process with rate $\lambda = 2$ packet/s. We are interested in estimating the amount of time T for the observation time interval $[0, T]$ that maximizes the probability of having exactly one packet arrival.

Solution

We have one packet arrival within $[0, T]$ with probability:

$$P(N(T) = 1) = \frac{(\lambda T)^1}{1!} e^{-\lambda T}$$

To obtain the value of T that maximizes this probability, we need to differentiate $P(N(T) = 1)$ with respect to T and set the result to zero:

$$\frac{d}{dT} P(N(T) = 1) = 0 = \lambda e^{-\lambda T} + \lambda T e^{-\lambda T}(-\lambda)$$

Solving the above equation brings:

$$T_{opt} = \frac{1}{\lambda}$$

which is the average value of an exponential random variable with a parameter λ. In conclusion, the best time range for exactly one arrival is the average inter-arrival time of events.
Remark that the result is a maximum and not a minimum if:

$$\left. \frac{d^2}{dT^2} P(N(T) = 1) \right|_{T = T_{opt}} < 0$$

Furthermore, when $T = T_{opt} = \frac{1}{\lambda}$, we have exactly one arrival with probability:

$$P(N(T=1)) = \lambda \frac{1}{\lambda} e^{-\lambda \frac{1}{\lambda}} = e^{-1} = 0.37$$

Problem 4

A cellular company wishes to install an antenna to cover a rural area that serves about 40 users. Consider that each user can be modeled by a Poisson process and make on average 1.5 phone calls per day. (a) Find the average number of phone calls produced in a random interval of one hour. (b) Find the probability of having no phone calls for one hour selected randomly. (c) Find the probability of exactly 1 or 2 phone calls in 15 minutes taken at random.

Solution

The aggregated process is Poisson with rate:

$$\lambda_{agg} = \sum_{i=1}^{40} \lambda_i = 40 \times 1.5 = 60 \text{ call/day} = 2.5 \text{ call/hour}$$

The average number of calls per hour is then:

$$E(N(1h)) = \lambda_{agg}T = 2.5 \text{ call/hour} \times 1 \text{ hour} = 2.5 \text{ calls}$$

There are no calls in one hour with probability:

$$P(N(1h) = 0) = \frac{(\lambda T)^0}{0!} e^{-\lambda T} = e^{-2.5 \cdot 1} = 0.08$$

In 15 mins, we have exactly one call with probability:

$$P(N(0.25h) = 1) = \frac{(\lambda T)^1}{1!} e^{-\lambda T} = 2.5 \times 0.25 e^{-2.5 \times 0.25} = 0.3345$$

and two calls with probability:

$$P(N(0.25h) = 2) = \frac{(\lambda T)^2}{2!} e^{-\lambda T} = \frac{(2.5 \times 0.25)^2}{2!} e^{-2.5 \times 0.25}$$
$$= 0.1045$$

Problem 5

An Ethernet switch with 24 input ports, $I_1, \ldots, I_{24}$ receives packets following a Poisson process with rate $\lambda_i = 10$ packet/s per port. This switch has got two output ports. Packets coming from the first 18 input ports go directly to the first output port, while packets coming from the other 6 input ports go to the second output port. Find: (a) The outgoing packet rate of the two output ports; (b) the probability that, at a given random time, the next two packets both go to the first output port; (c) the probability that exactly 20 packet arrivals for the first output port occur in the next 20 milliseconds; (d) the probability that exactly 30 packets arrive at the switch within 10 ms, and that 20 of them go to the first output port.

Solution

First of all, the packet process at the two output ports are Poisson since they are the aggregate of several individual Poisson processes. The two output ports o_1 and o_2 have the following outgoing rates:

$$\lambda_{o_1} = 18 \times \lambda_1 = 180 \text{ packet/s}$$
$$\lambda_{o_2} = 6 \times \lambda_1 = 60 \text{ packet/s}$$

If a packet is selected at random, then this packet goes to the first output port with probability $\frac{18}{24}$, and to the second port with probability $\frac{6}{24}$. We may have four cases for two consecutive packets:

(First port, First port) with prob: $\frac{18}{24} \times \frac{18}{24}$

(First port, Second port) with prob: $\frac{18}{24} \times \frac{6}{24}$

Second port, First port) with prob: $\frac{6}{24} \times \frac{18}{24}$

(Second port, Second port) with prob: $\frac{6}{24} \times \frac{6}{24}$

Hence, two consecutive packets go to the same output port with probability:

$$\frac{18}{24} \times \frac{18}{24} + \frac{6}{24} \times \frac{6}{24} = \frac{5}{8}$$

The probability that exactly 20 packets arrive for the first output port in the time interval $[0, 20ms]$ follows:

$$P(N_{o_1}(T = 20ms) = 20 \text{ packets}) = \frac{(180 \times 0.02)^{20}}{20!} e^{-180 \times 0.02}$$
$$= 1.5 \cdot 10^{-9}$$

The probability that exactly 30 packets arrive at the switch and that 20 of them go to the first output port:

$$P(N_{o_1}(T = 20ms) = 20 \text{ packets})$$
$$\times \ P(N_{o_2}(T = 20ms) = 10 \text{ packets})$$

since this condition requires 20 packet arrivals for the first output port and 10 packet arrivals for the second output port. Solving:

$$\frac{(180 \times 0.02)^{20}}{20!} e^{-180 \times 0.02} \times \frac{(60 \times 0.02)^{10}}{10!} e^{-60 \times 0.02} = 7.7 \cdot 10^{-16}$$

Alternatively, we may compute the probability of having exactly 30 packets on both ports as:

$$P(N_{o_1+o_2}(T = 20ms) = 20 \text{ packets}) = \frac{(240 \times 0.02)^{30}}{30!} e^{-240 \times 0.02}$$

and then multiply this result by the probability that exactly 20 of them are for the first output port (binomial distribution):

$$\binom{30}{20} \left(\frac{3}{4}\right)^{20} \left(1 - \frac{3}{4}\right)^{30-20}$$

which yields the same result.

Problem 6

Consider two football teams playing a game. We have observed that the goals scored by each team follow a Poisson process with rates $\lambda_1 = 0.02$ and $\lambda_2 = 0.015$ goal/min. Consider that you have 1 euro to bet on a score and that you are given two options: 2-1 with a 10 euro benefit, or 2-2 with a 20 euro benefit. Which score provides a better average benefit? Assume that a whole game lasts for 90 minutes exactly.

Solution

First, we need to compute the probability of each result. First, score 2-1 occurs with probability:

$$\begin{aligned}
P(\text{Score 2-1}) &= P(N_1(t) = 2, N_2(t) = 1) \\
&= \frac{(\lambda_1 T)^2}{2!} e^{-\lambda_1 T} \times \frac{(\lambda_2 T)^1}{1!} e^{-\lambda_2 T} \\
&= \frac{(0.02 \cdot 90)^2}{2!} e^{-0.02 \cdot 90} \times \frac{(0.015 \cdot 90)^1}{1!} e^{-0.015 \cdot 90} \\
&= 0.27 \times 0.35 = 0.0945
\end{aligned}$$

Score 2-2 occurs with probability:

$$P(\text{Score 2-2}) = P(N_1(t) = 2, N_2(t) = 2)$$
$$= \frac{(0.02 \cdot 90)^2}{2!} e^{-0.02 \cdot 90} \times \frac{(0.015 \cdot 90)^2}{2!} e^{-0.015 \cdot 90}$$
$$= 0.27 \times 0.24 = 0.065$$

The average benefit from the first score is:

$$E(\text{Benef}) = 10 \times P(\text{Score 2-1}) + 0 \times (1 - P(\text{Score 2-1}))$$
$$= 10 \times 0.0945 = 0.945 \text{ euro}$$

and:

$$E(\text{Benef}) = 20 \times P(\text{Score 2-2}) + 0 \times (1 - P(\text{Score 2-2}))$$
$$= 20 \times 0.065 = 1.3 \text{ euro}$$

for the second result.
As shown, the first result is more likely to occur than the second one. However, the first result provides an average benefit of 0.945 euros per euro bet, a loss rather than a benefit.

Problem 7

Consider an autonomous vehicle scenario with $V = 10$ vehicles and two computing servers, server A and server B. Each vehicle $v = 1, 2, \ldots, V$ produces computing tasks following a Poisson distribution with rate $\lambda_v = 5$ tasks/sec, and directs its tasks to server A with probability $p = 0.2$ and to server B with probability $(1 - p) = 0.8$.

1. what is the rate with which tasks arrive at each of the servers?

2. what is the average interarrival time of tasks at each server?

3. what is the average time that the 20-th task is produced? The average time of the 20-th task at each of

the servers?

4. what is the probability that exactly 45 tasks arrived at server A from second $s_1 = 0$ to second $s_2 = 5$? How and why does the answer change in the next 5 seconds, i.e. from second $s_2 = 5$ to second $s_3 = 10$?

Solution

1. The process that describes the generation of tasks is the aggregation of the individual Poisson processes of all vehicles, being itself a Poisson with Parameter $\lambda = \sum_{v=1}^{V} \lambda_v = 50$ tasks/sec. The process that describes the arrival of tasks at each server is a sampling of the original one, with sampling probability p and $(1 - p)$ for server A and B, respectively. Thus, it is a Poisson, with parameters:

$$\lambda_A = V p \lambda = 10 \cdot 0.2 \cdot 5 = 10 \text{task/sec},$$
$$\lambda_B = V(1-p)\lambda = 10 \cdot 0.8 \cdot 5 = 40 \text{task/sec}.$$

2. Let X_{iA} and X_{iB} be the inter-arrival time of task i at server A and B, respectively. Then, given that the arrival processes at the two servers are Poisson processes, it is:

$$E(X_{iA}) = \frac{1}{\lambda_A} = \frac{1}{10} = 0.1 \ sec,$$
$$E(X_{iB}) = \frac{1}{\lambda_B} = \frac{1}{40} = 0.025 \ sec.$$

3. The absolute arrival time S_i of task i follows the sum of exponentially distributed r.v. $X_j, j = 0,...i$ (since the processes are Poisson). This sum is known to follow the Gamma distribution with same parameter λ as the Poisson, and i degrees of freedom, with the following mean and variance:

$$E(S_i) = \frac{i}{\lambda}, \ V(S_i) = \frac{i}{\lambda^2}.$$

Thus:
The average time that the 20-th task is produced is:

$$E(S_i) = \frac{i}{\lambda} = \frac{20}{50} = 0.4 sec,$$

$$Std(S_i) = \sqrt{\frac{i}{\lambda^2}} = \sqrt{\frac{20}{50^2}} = 0.089 sec.$$

So the 20-th event will most likely lie within 0.4 ± 0.178 sec.
For the 20-th task arriving at server A:

$$E(S_{iA}) = \frac{i}{\lambda_A} = \frac{20}{10} = 2 sec,$$

$$Std(S_{iA}) = \sqrt{\frac{i}{\lambda_A^2}} = \sqrt{\frac{20}{10^2}} = 0.447 sec.$$

So the 20-th event will most likely lie within 2 ± 0.447 sec.
Similarly, for the 20-th task arriving at server B:

$$E(S_{iB}) = \frac{i}{\lambda_B} = \frac{20}{40} = 0.5 sec,$$

$$Std(S_{iB}) = \sqrt{\frac{i}{\lambda_B^2}} = \sqrt{\frac{20}{40^2}} = 0.111 sec.$$

So the 20-th event will most likely lie within 0.5 ± 0.111 sec.

4. Let $N(t)$ be the number of tasks that are produced from second 0 to second t. Since the process of arrival of tasks at server A is a Poisson process with parameter $\lambda_A = 10$ task/sec, the number of events occurred within a time interval of length t follows the Poisson distribution with rate $\lambda_A t$. Thus:

$$P(N(t) = n) = \frac{(\lambda_A t)^n}{n!} e^{-\lambda_A t} = \frac{(10 \cdot 5)^{45}}{45!} e^{-10 \cdot 5} = 0.046. \quad (3.9)$$

For the interval [5sec, 10sec] the answer does not change, as the Poisson processes have stationary increments.

Problem 8

1. Consider a rural area with 1,000 temperature sensors deployed around the 100 km². Each sensor sends messages to a WiFi access point at a pace of 10 messages per minute; inter-departure times of messages are uniformly distributed between $[0, b]$

seconds (for each sensor). Can you find the value of b?

2. Can you model the arrival process of messages at the WiFi access points which contains all the messages from all 1,000 sensors using a Poisson distribution? Why? What is the value of λ for this process?

3. Can you find the average number of message arrivals in 1 second and 3 minutes? And the variance? Can you find the moment at which message number 200 arrives at the access point (average and variance). Tip: you can use the Central Limit Theorem.

Solution

If a temperature sensor sends messages at a rate of 10 messages per minute (or 60 seconds), this means that, on average, one message is output from the sensor every 6 seconds. We know that the time between output messages from the sensor is random according to a uniform distribution $U(0, b)$. That is, the average time between two messages is:

$$\frac{a+b}{2} = \frac{0+b}{2} = 6 \ s$$

Therefore, $b = 12 \ sec$. That is, the time between two messages is uniform between 0 and 12 seconds.

According to the Palm-Khintchine theorem, it can be assured that the aggregate process of arrival of messages (coming from 1000 sensors) to the WiFi station would approximate a Poisson Process, since there are many independent processes (specifically 1000) of similar intensity. The rate is:

$$\lambda_{agg} = 1000 \cdot 10 \ mes/min = 10000 \ mes/min = 166.67 \ mes/seg$$

Both the mean and variance in the Poison process are:
$$E(N(t)) = Var(N(t)) = \lambda t$$
Thus:
$$E(N(t = 1s)) = Var(N(t = 1s)) = \lambda t = 166.67 \text{ mes/s} \cdot 1s = 166.67 \text{ mes}$$

$$E(N(t = 180s)) = \lambda t = 166.67 \text{ mes/s} \cdot 180s = 30000 \text{ mes}$$
Also:
$$E(N(t = 180s)) = Var(N(t = 180s)) = 30000 \text{ mes}$$

In the aggregated process, the time between messages arriving at the WiFi station follows an exponential distribution of mean and standard deviation equal to $1/\lambda = 0.006$ sec. The sum of 200 iid random variables with mean and standard deviation equal to 0.006 s is distributed according to a normal with the following parameters:
$$N(200\mu, 200\sigma^2)$$
That is, the 200-th message arrives at the sensor at time $200 \cdot 0.006 = 1.2$ sec on average, and its standard deviation is
$$\sqrt{200 \cdot 0.006^2} = 0.085 \text{ sec}$$

4
Discrete-Time Markov Chains

4.1 Introduction by example

Consider a truck driver conducting his business in three cities: London, Oxford, and Nottingham. We assume that the truck driver carries products from one city to another, but performs only one trip per day. For example, consider that, on day 0, the truck driver departs from London with products destined for Oxford. On day 1, the truck driver is in Oxford, and the destination of the products he collects in Oxford may be different than those collected in London. In other words, the next city of the truck driver depends only on the actual city from which he departs.

Table 4.1 gives the probability values to go from one city to the next. As shown, when the truck driver is in London, the probability of delivering a product with destination Oxford is much higher (0.8) than the probability of going to Nottingham (0.2). In Oxford, the truck driver is more likely to go back to London (0.6), rather than going to Nottingham (0.3). In addition, there is some probability (0.1) of going from somewhere in Oxford to somewhere else in Oxford (internal delivery). Finally, when the truck driver departs from Nottingham, he is very likely to do internal delivery in Nottingham (0.6), but may also go to London (0.2) or Oxford (0.2).

From-To	Lond.	Oxf.	Nott.
London	0	0.8	0.2
Oxford	0.6	0.1	0.3
Nott.	0.2	0.2	0.6

Table 4.1: Source-destination probabilities

This is the classical example where Discrete-Time Markov Chains (DTMCs) perfectly fit. The following questions about the truck driver can be easily addressed with the help of DTMCs:

1. Assuming that the truck driver starts in London on day 0, where is he most likely going to be on day 3? To answer this question, we first need to introduce *the Chapman-Kolmogorov equations*.

2. Assuming that the truck driver reaches Nottingham on a given day, for how many consecutive days will he stay in Nottingham on average? This is the concept of *sojourn times*.

3. What percentage of days will the truck driver spend in each city in a year? This is solved by studying the *steady-state distribution* of the DTMC.

4. Assuming that the truck driver starts in London on the first day, when is he expected to reach the city of Nottingham for the first time? This is solved by applying the *first-passage time* concepts of DTMCs.

4.2 States and transitions between states

In every DTMC problem, the first important task is to identify the chain states. The states represent the different situations that the chain may take.

In the truck driver example, we have three states that relate to the city in which the truck driver is on a given day:

London, Oxford, and Nottingham. Once the truck driver has reached a state, say London, he may jump to a different state on the next day, for instance, Oxford.

The next figure shows the state-transition diagram, which provides a visual summary of the DTMC. In this figure, the circles refer to the three states, and the arcs represent the transitions between states with their respective probabilities. This figure completely characterizes the DTMC.

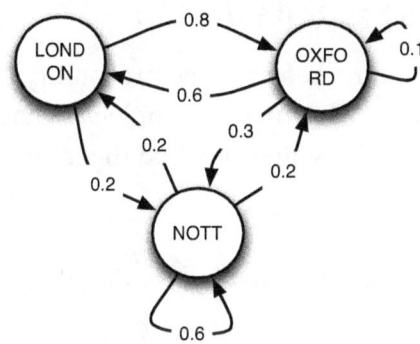

Figure 4.1: State-transition diagram

Now, let X_0 refer to the state at which the chain is on day 0, for example, $X_0 = London$. The chain then jumps to another state (possibly different) on day 1, denoted by X_1, for example $X_1 = Oxford$. From this new state, we consider that the chain jumps to another state on day 2, for example, $X_2 = London$ again, and so on. This sequence $\{London, Oxford, London, ...\}$ comprises *a path* of the DTMC. A path is a possible realization of the process defined by the DTMC. Other possible paths of this chain (assuming the truck driver starts in London) are:

$$\{London, Oxford, Oxford, Nott, Nott, Oxford, ...\}$$

$$\{London, Nott, Oxford, Nott, Nott, Nott, ...\}$$

Finally, we can also collect the transition state probabilities into the following *transition matrix P*:

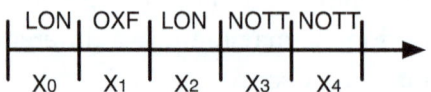

Figure 4.2: Example of a path.

$$P = \begin{pmatrix} 0 & 0.8 & 0.2 \\ 0.6 & 0.1 & 0.3 \\ 0.2 & 0.2 & 0.6 \end{pmatrix} \quad (4.1)$$

where each item p_{ij} in the matrix refers to the probability of jumping from state i to state j. This matrix completely characterizes the DTMC, just like the state-transition diagram.

> **Example 1**
>
> Which of the following paths are possible for the truck driver example?
>
> 1. $\{London, London, Oxford, Nott, Nott, Oxford\}$
> 2. $\{London, Oxford, Oxford, Oxford, Oxford, Nott\}$

> **Solution**
>
> The first path is not possible since this Markov chain does not allow a transition from London to London (probability 0). The second path is possible, although very unlikely since the transitions from Oxford to Oxford occur with a very small probability (only 0.1).

4.3 Problem formulation

Let X_n denote the city at which the truck driver is on day n. The subindex n in X_n denotes the time slot in the discrete chain, namely $n = 0, 1, 2, 3, \ldots$ In the truck driver example, every time slot is of length one day.

For simplicity, let us consider: London as state 1, Oxford as state 2, and Nottingham as state 3. In this case, the *sample space* for this chain is $S = \{1, 2, 3\}$, and the chain may take any of these values $X_n \in \{1, 2, 3\}$ on a given day $n = 0, 1, 2, \ldots$.

Day 0 ($n = 0$) refers to the *initial state* of the chain. For instance, $X_0 = 1$ refers to the case where the truck driver starts his business in the city of London. Of course, the chain may start on any other state. If we consider that the truck driver may start in any city with the same probability, then we would say:

$$P(X_0 = 1) = P(X_0 = 2) = P(X_0 = 3) = \frac{1}{3}$$

However, if we are told that the truck driver always starts his business in the city of London, then:

$$P(X_0 = 1) = 1$$
$$P(X_0 = 2) = 0$$
$$P(X_0 = 3) = 0$$

In summary, we may collect this information into the so-called *vector of initial probabilities* $\pi^{(0)}$:

$$\pi^{(0)} = \begin{pmatrix} \pi_1^{(0)} & \pi_2^{(0)} & \pi_3^{(0)} \end{pmatrix} \quad (4.2)$$

where the i-th item in the vector refers to the probability of having the truck driver starting in the i-th city, i.e.:

$$P(X_0 = i) = \pi_i^{(0)}$$

Obviously:

$$\sum_{i=1}^{3} \pi_i^{(0)} = 1 \quad (4.3)$$

Example 2

Consider the following initial probability vectors:

1. $\pi^{(0)} = \begin{pmatrix} 0.5 & 0.5 & 0 \end{pmatrix}$

2. $\pi^{(0)} = \begin{pmatrix} 0.25 & 0.5 & 0.25 \end{pmatrix}$

3. $\pi^{(0)} = \begin{pmatrix} 0.25 & 0 & 0.75 \end{pmatrix}$

4. $\pi^{(0)} = \begin{pmatrix} 0.5 & 0 & 0.5 \end{pmatrix}$

Which one refers to the case where the truck driver may only start from either London or Nottingham with the same probability?

Solution

The correct answer is:
$$\pi^{(0)} = \begin{pmatrix} 0.5 & 0 & 0.5 \end{pmatrix}$$

For now, let us consider that the truck driver starts his business from the city of London, then:
$$P(X_0 = 1) = 1$$
and
$$P(X_0 = 2) = P(X_0 = 3) = 0$$

The next step is to guess the driver's city on day 1, that is, to obtain $P(X_1 = i)$, for $i = 1, 2, 3$. In other words, this is to obtain the probability that the driver will be in London, Oxford or Nottingham on day 1. From the state-transition diagram, we observe that:
$$P(X_1 = 1) = 0, \quad P(X_1 = 2) = 0.8, \quad P(X_1 = 3) = 0.2$$
since the truck driver cannot go from London to London, may go from London to Oxford (with probability 0.8) or Nottingham (with probability 0.2).

Now, consider that the initial state is not known, and we decide to assume that all states are equally likely, that is:
$$\pi^{(0)} = \begin{pmatrix} \frac{1}{3} & \frac{1}{3} & \frac{1}{3} \end{pmatrix}$$
Then, the probability values $P(X_1 = i)$ change completely.

For instance, the probability of being in London on day 1 has to take into account all the possible cases of day 0. Using conditioning, this is:

$$P(X_1 = 1) = \sum_{i=1}^{3} P(X_1 = 1 | X_0 = i) P(X_0 = i)$$
$$= \pi_1^{(0)} p_{11} + \pi_2^{(0)} p_{21} + \pi_3^{(0)} p_{31}$$

Here, the term $p_{ij} = P(X_n = j | X_{n-1} = i)$ refers to the transition probability from state i to state j. The previous equation shows the fact that *the state on day 1 depends on the state of the previous day*.

Solving, we obtain:

$$P(X_1 = 1) = \frac{1}{3} 0 + \frac{1}{3} 0.6 + \frac{1}{3} 0.2 = 0.267$$

Similarly, we can obtain the probability of being in Oxford and Nottingham on day 1. The solution is: $P(X_1 = 2) = 0.367$ and $P(X_1 = 3) = 0.367$.

We can now collect these probability values into a vector for day 1 as:

$$\pi^{(1)} = \begin{pmatrix} 0.267 & 0.367 & 0.367 \end{pmatrix}$$

It is worth noting that these three values again add up to 1.

In general, it can be observed that the value of $\pi^{(1)}$ can also be obtained using the transition probability matrix as:

$$\pi^{(1)} = \pi^{(0)} P$$

4.4 The Markov property and the Chapman-Kolmogorov equations

To obtain vector $\pi^{(2)}$, we need to introduce the Markov property of DTMCs:

Markov property Let $(X_0, X_1, \ldots)$ be a Markov chain with state space $S = \{s_1, \ldots, s_K\}$ and transition matrix P. The Markov property of DTMCs states that the actual state of the chain only depends on the previous state, not on the whole previous set of states. In other words:

$$P(X_n = s_1 | X_{n-1} = s_2, \ldots, X_0 = s_{n+1}) = P(X_n = s_1 | X_{n-1} = s_2), \quad s_l \in S \quad (4.4)$$

Thanks to the Markov property and conditioning, we can easily formulate the city at which the truck driver is on day 2 as:

$$P(X_2 = j) = \sum_{i=1}^{3} P(X_2 = j | X_1 = i) P(X_1 = i)$$

$$= \sum_{i=1}^{3} p_{ij} \pi_i^{(1)}, \quad j = 1, 2, 3 \quad (4.5)$$

This equation is often referred to as *the Chapman-Kolmogorov equation*. In a matrix form, we can summarise the above equation as:

$$\pi^{(2)} = \pi^{(1)} P \quad (4.6)$$

It can be further noted that:

$$\pi^{(2)} = \pi^{(1)} P = (\pi^{(0)} P) P = \pi^{(0)} P^2 \quad (4.7)$$

The same procedure can be used to obtain the row vectors $\pi^{(3)}$, $\pi^{(4)}$, etc. In general:

$$\pi^{(n)} = \pi^{(n-1)} P = \pi^{(0)} P^n, \quad n \geq 1 \quad (4.8)$$

> ### Example 3
>
> Assuming that the truck driver starts in London on day 0, where is he most likely going to be on day 3?
>
> #### Solution
>
> To answer this question we just need to apply the Chapman-Kolmogorov equations:
>
> $$\pi^{(3)} = \pi^{(0)} P^3$$
>
> where: $\pi^{(0)} = \begin{pmatrix} 1 & 0 & 0 \end{pmatrix}$ Thus:
>
> $$\pi^{(3)} = \begin{pmatrix} 0.14 & 0.5 & 0.36 \end{pmatrix}$$

Hence the truck driver will mostly likely be in Oxford (with probability 0.5).

4.5 Sojourn times

Now, consider that the truck driver reaches the state Oxford at a given time slot n. We would like to know how many consecutive time slots the truck driver will stay on average in that state.

Let D_2 refer to the random variable describing the number of consecutive time slots that the truck driver spends in state 2 (Oxford) once he reaches it. Essentially, when the truck driver reaches Oxford, he may leave Oxford in the next day with the following probability:

$$1 - p_{22} \text{ for path: } \{\ldots, Oxf, out\}$$

where *out* refers to the other two states: London or Nottingham. Alternatively, we may have the following path:

$$\{\ldots, Oxf, Oxf, out\} \text{ which occurs with probability } p_{22}(1 - p_{22})$$

or path:

$$\{\ldots, Oxf, Oxf, Oxf, out\} \text{ which occurs with probability } p_{22}^2(1 - p_{22})$$

etc.

In conclusion, the probability of spending exactly d time slots in Oxford, once the truck driver has just reached it, follows:

$$P(D_2 = d) = p_{22}^{d-1}(1 - p_{22}), \quad d = 1, 2, \ldots \quad (4.9)$$

This probability distribution is the well-known geometric distribution and has the following mean:

$$E(D_2) = \frac{1}{1 - p_{22}} \quad (4.10)$$

Example 4

Assuming that the truck driver reaches Nottingham on a given day, (1) find the probability of staying exactly 2 days in Nottingham; and (2) the average number of consecutive days spent in Nottingham, London, and Oxford.

Solution

The truck driver spends exactly 2 days in Nottingham with probability:

$$P(D_{Nott} = 2) = p_{33}(1 - p_{33}) = 0.6(1 - 0.6) = 0.24$$

which refers to the probability of jumping from Nottingham to Nottingham, and then from Nottingham to either London or Oxford, thus exactly two days in Nottingham.

We know that the sojourn time of state i is geometrically distributed with parameter p_{ii}. Hence, the average number of consecutive days in Nottingham follows eq. 4.10:

$$E(D_{Nott}) = \frac{1}{1 - p_{33}} = \frac{1}{1 - 0.6} = 2.5 \text{ time days}$$

The average sojourn time for London is exactly one day as noted from the:

$$E(D_{Lon}) = \frac{1}{1 - p_{11}} = \frac{1}{1 - 0} = 1 \text{ time day}$$

since once the chain reaches that state, it automatically leaves it on the next time slot. Finally, the sojourn time for Oxford is:

$$E(D_{Oxf}) = \frac{1}{1 - p_{22}} = \frac{1}{1 - 0.1} = 1.11 \text{ time days}$$

4.6 Reachability and types of states

A state s_i communicates with another state s_j (denoted by $s_i \to s_j$) if the probability of reaching s_j when starting from s_i is strictly positive. In such a case, state s_j is said to be *accessible* from s_i. Mathematically:

$$s_i \to s_j \quad \text{if } P(X_{n+m} = s_j | X_n = s_i) > 0 \text{ for some } m > 0 \quad (4.11)$$

It can be further shown that such probability equals $(P^m)_{ij}$, that is, the (i,j)-th position of matrix P^m.

If $s_i \to s_j$ and $s_i \leftarrow s_j$, then we may say that the states s_i and s_j *intercommunicate* with each other, and we write $s_i \leftrightarrow s_j$.

A Markov chain is said to be *irreducible* if all its states intercommunicate with each other. Otherwise, the chain is said to be *reducible*.

A state s_i is said to be *periodic* and has period m if any return to state s_i must occur in multiples of m time steps; otherwise the state is said to be *aperiodic* (returns to state s_i can occur at irregular times). Finally, a Markov chain is aperiodic if all its states are aperiodic.

A state s_i is called *recurrent* if, with probability 1, the process will eventually return to that state after it leaves it.

A state s_i is called *absorbing* if it is impossible to leave that state once the chain has reached it, i.e. $p_{ii} = 1$.

> ### Example 5
>
> The next figure shows three Markov Chains. What can you say about their states?
>
>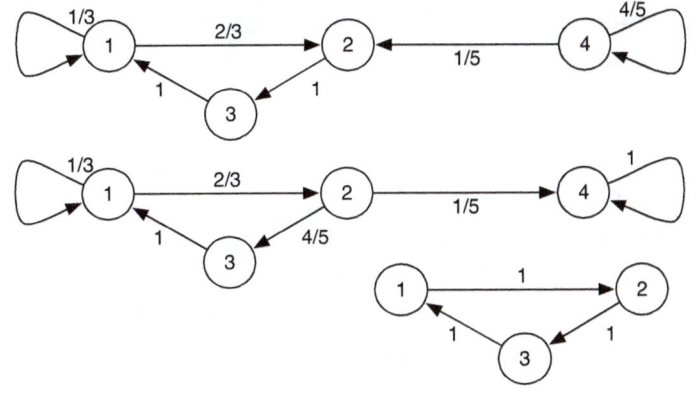

> **Solution**
>
> In the first Markov chain, states 1, 2, and 3 are recurrent since the process will eventually return to these states. However, state 4 is not recurrent since, once the chain leaves it, it may never return to it. State 4 is not accessible from states 1, 2, and 3.
>
> In the second Markov chain, state 4 is now absorbing: Once the chain reaches state 4, it is trapped in that state.
>
> Finally, the third Markov chain is periodic.

4.7 Steady-state analysis

Let $(X_0, X_1, \ldots)$ be a Markov chain with state space $S = \{s_1, \ldots, s_K\}$ and transition matrix P. A row vector π is said to be a *stationary distribution* for the Markov chain if it satisfies:

1. $\pi_i \geq 0$ for $i = 1, 2, \ldots$ and $\sum_{i=1}^{K} \pi_i = 1$

2. $\pi P = \pi$, in other words: $\sum_{i=1}^{K} \pi_i p_{ij} = \pi_j$

Other terms used for vector π are: *invariant distribution*, emphequilibrium distribution or *limiting-state probabilities*.

It can be proved that any irreducible and aperiodic chain has exactly one stationary distribution π. It can be further proved that this stationary distribution π satisfies:

$$\lim_{n \to \infty} \pi^{(n)} = \pi$$

that is, the stationary distribution also provides the probability that the chain is in any state after a (long) number of time slots. This means that the stationary distribution gives the percentage of time that the Markov Chain spends in each state in the long run (after a long number of state transitions).

Example 6

We can now answer the third question about the truck driver: What percentage of days will the truck driver spend in each city in a year?

Solution

To answer this question, we need to apply the steady-state analysis of DTMCs for the truck driver example. Essentially, we need to find the vector π that satisfies:

$$\pi P = \pi \quad \text{with} \quad \sum_i \pi_i = 1$$

Thus:

$$\begin{pmatrix} \pi_1 & \pi_2 & \pi_3 \end{pmatrix} \begin{pmatrix} 0 & 0.8 & 0.2 \\ 0.6 & 0.1 & 0.3 \\ 0.2 & 0.2 & 0.6 \end{pmatrix} = \begin{pmatrix} \pi_1 & \pi_2 & \pi_3 \end{pmatrix}$$

which produces the following set of equations:

$$0.6\pi_2 + 0.2\pi_3 = \pi_1$$
$$0.6\pi_1 + 0.1\pi_2 + 0.2\pi_3 = \pi_2$$
$$0.2\pi_1 + 0.3\pi_2 + 0.6\pi_3 = \pi_3$$
$$\pi_1 + \pi_2 + \pi_3 = 1$$

The first three questions are linearly dependent so we may remove one of them, for instance, the second one. This gives a set of three equations with three unknowns:

$$0.6\pi_2 + 0.2\pi_3 = \pi_1$$
$$0.2\pi_1 + 0.3\pi_2 + 0.6\pi_3 = \pi_3$$
$$\pi_1 + \pi_2 + \pi_3 = 1$$

Solving:

$$\pi_1 = 0.28 \quad \pi_2 = 0.33 \quad \pi_3 = 0.39$$

Hence, over a long period, say one year, the truck driver spends 28% of the days in London, 33% in Oxford and 39% in Nottingham.

4.8 First-passage times

Finally, to answer the last question about the truck driver, we need to introduce the concept of *first-passage times*. Let m_{ij} refer to the mean first passage time from state i to state j. Then, it can be shown that:

$$m_{ij} = 1 + \sum_{k \neq j} p_{ik} m_{kj}, \text{ with } i \neq j \qquad (4.12)$$

which arises after applying the properties of conditional expectation.

The meaning of this equation is as follows: The average number of time slots required to go from state i to state j equals to one time slot plus the average first passage time from any other intermediate state k ($k \neq j$) to state j multiplied by the probability that the next intermediate state is state k.

Similarly the mean recurrence time m_{ii} can be computed as:

$$m_{ii} = 1 + \sum_{k \neq i} p_{ik} m_{ki} \qquad (4.13)$$

Example 7

Assuming that the truck driver starts in London on the first day when is he expected to reach the city of Nottingham for the first time?

Solution

The question involves computing the value of m_{13}, that is, the average first-passage time from London to Nottingham. Hence, using Eq. 4.12 we observe that:

$$m_{13} = 1 + p_{11}m_{13} + p_{12}m_{23} = 1 + 0.8m_{23}$$

This equation has two unknowns: m_{13} and m_{23}. We then need to apply Eq. 4.12 again to obtain m_{23}:

$$m_{23} = 1 + p_{21}m_{13} + p_{22}m_{23} = 1 + 0.6m_{13} + 0.1m_{23}$$

Now, we have two equations with two unknowns. Solving, we obtain:
$$m_{13} = 4.05 \text{ days}, \quad m_{23} = 3.81 \text{ days}$$

4.9 Further problems

Problem 1

Verify that the following Markov chain is irreducible, and obtain the steady-state vector.

$$P = \begin{pmatrix} 1/2 & 1/2 & 0 \\ 1/2 & 1/4 & 1/4 \\ 0 & 1/3 & 2/3 \end{pmatrix}$$

Solution

After plotting its state-transition diagram (see next figure), we observe that all states intercommunicate with each other. Hence, the chain is irreducible, and we can obtain its steady-state vector π.

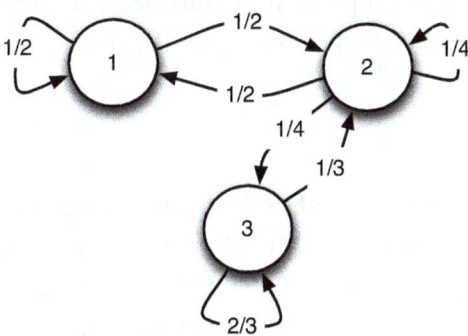

To do so, we need to solve:

$$\pi P = \pi \quad \text{with} \quad \sum_i \pi_i = 1$$

which gives the following system of equations:

$$\frac{1}{2}\pi_1 + \frac{1}{2}\pi_2 = \pi_1$$
$$\frac{1}{2}\pi_1 + \frac{1}{4}\pi_2 + \frac{1}{4}\pi_3 = \pi_2$$
$$\frac{1}{3}\pi_2 + \frac{2}{3}\pi_3 = \pi_3$$
$$\pi_1 + \pi_2 + \pi_3 = 1$$

Since we only need two of the first three equations, we may remove the second one. After solving, we obtain:

$$\pi_1 = 0.36 \quad \pi_2 = 0.36 \quad \pi_3 = 0.28$$

Problem 2

Consider a DTMC with the following transition matrix:

$$P = \begin{pmatrix} 1/2 & 1/2 \\ 0 & 1 \end{pmatrix}$$

Draw the transition graph of this Markov chain and show whether or not this chain is irreducible

Solution

The next figure shows the state-transition diagram of this DTMC, which is characterized by matrix P. As observed, state 2 is an absorbing or trapping state, hence the Markov chain is not irreducible since not all states intercommunicate.

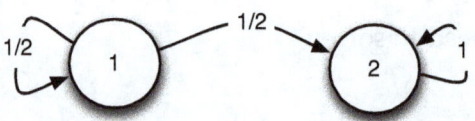

Problem 3

Three out of every four voice packets in a communications link are followed by a data packet, while only one out of every ten data packets are followed by a voice packet. What fraction of transmitted packets are data packets?

Solution

We can use a two-state DTMC to model this problem. The two states refer to whether a transmitted packet is of Data or Voice type. In addition, the transition probabilities from Voice to Data and vice versa are also stated in the problem as conditional probabilities from one state to the next. This yields the next DTMC model.

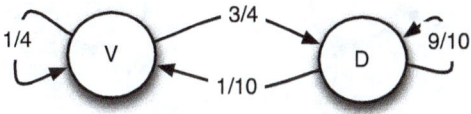

Solving for the steady state, i.e.:

$$\pi P = \pi \quad \text{with} \quad \sum_i \pi_i = 1$$

we obtain:

$$\pi_V = 0.12 \quad \pi_D = 0.88$$

Problem 4

Consider a three-node token-based network with slotted time. In this network, token possession gives the right to transmit data over the ring in a given time slot. We assume that both the token and the data circulate in the clockwise direction. Consider that, when the first node has the token, it decides to keep the token for the next round with a probability of 0.5. The second and third nodes keep the token for the next time slot with probability 0.2 and 0.7 respectively. Initially at time-slot zero, node 1 has the token. (a) Find the probability that the second and third nodes will have the token in the first, second, and third time slot; (b) obtain the stationary distribution for this DTMC. (c) Compute the probability that each node possesses the token at time slots 20 and 100.

Solution

This problem can be modeled with a three-state DTMC. Each state identifies which node in the ring currently holds the token. The next figure shows the state-transition diagram for the token ring.

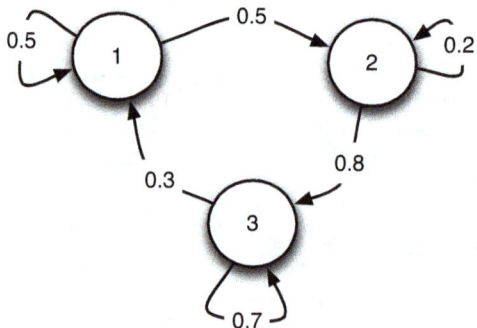

The transition matrix P is then:
$$P = \begin{pmatrix} 0.5 & 0.5 & 0 \\ 0 & 0.2 & 0.8 \\ 0.3 & 0 & 0.7 \end{pmatrix}$$

We also know the initial distribution vector:
$$\pi^{(1)} = \begin{pmatrix} 1 & 0 & 0 \end{pmatrix}$$

Hence:
$$\pi^{(2)} = \pi^{(1)} P = \begin{pmatrix} 0.5 & 0.5 & 0 \end{pmatrix}$$
$$\pi^{(3)} = \pi^{(2)} P = \begin{pmatrix} 0.25 & 0.35 & 0.4 \end{pmatrix}$$

The stationary distribution comes after solving:
$$\pi P = \pi \quad \text{with} \quad \sum_i \pi_i = 1$$
which yields:
$$\pi_1 = 0.30 \quad \pi_2 = 0.19 \quad \pi_3 = 0.51$$
Finally, we can approximate $\pi^{(20)} \approx \pi$ and $\pi^{(100)} \approx \pi$ since the chain is small (only three states) and it converges quickly to the steady-state vector.

Problem 5

In the token-based network above, obtain the average sojourn times for each node in the ring, and the average first passage time from node 1 to node 2, and from node 1 to node 3.

Solution

The average sojourn times for each state are:
$$E(D_1) = \frac{1}{1 - p_{11}} = \frac{1}{1 - 0.5} = 2 \text{ slots}$$
$$E(D_2) = \frac{1}{1 - p_{22}} = \frac{1}{1 - 0.2} = 1.25 \text{ slots}$$
$$E(D_3) = \frac{1}{1 - p_{33}} = \frac{1}{1 - 0.7} = 3.33 \text{ slots}$$

Clearly, the first passage time from state 1 to state 3 has to be:
$$E(D_1) + E(D_2) = 2 + 1.25 = 3.25 \text{ slots}$$
since the chain always go from state 1 to state 2 (and it takes 2 slots on average for that step) and then from state 2 to state 3 (taking 1.25 slots on average).

We can further prove this intuition by applying:
$$m_{13} = 1 + p_{11} m_{13} + p_{12} m_{23} = 1 + 0.5 m_{13} + 0.5 m_{23}$$
while:
$$m_{23} = 1 + p_{22} m_{23} = 1 + 0.2 m_{23}$$

Solving the second equation, we obtain: $m_{23} = \frac{1}{1-0.2} = 1.25$ slots. Using this information in the first equation, we obtain:

$$m_{13} = \frac{1 + 0.5 \cdot m_{23}}{1 - 0.5} = \frac{1 + 0.5 \cdot 1.25}{1 - 0.5} = 3.25 \text{ slots}$$

which gives the same result obtained previously.

Problem 6

In a slotted-time communications system, we have two stations sharing a common channel, and a token that is passed between the stations which grants access to the channel. The rules for exchanging/keeping the token are the following: When station 1 has got the token, it keeps it for the next time slot with probability p. Otherwise, it passes the token to station 2. When station 2 receives the token from station 1, it always keeps it for the next round, and then it returns the token to station 1. Essentially, station 1 may have the token for a variable number of time slots, but station 2 always has the token for two-time slots, and then it must return the token to station 1. (a) Use a DTMC to model this problem. Draw the state transition diagram and formulate the transition probability matrix. (b) Find the percentage of time that each station has got the token. (c) Find the value of p that guarantees that station 1 has got the token double that of station 2 on average. (d) Find the value of p that guarantees that each station holds the token with equal probability in the long run.

Solution

We can model this problem with a DTMC with three states: 1, 2 and 22. The first state considers the case that station 1 has the token and the other two states refer to the case when station 2 has got the token for the first time and for the second time respectively. The state transition diagram is shown in the next figure.

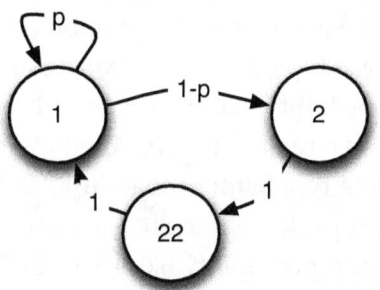

The transition probability matrix P depends on the value of p:

$$P = \begin{pmatrix} p & 1-p & 0 \\ 0 & 0 & 1 \\ 1 & 0 & 0 \end{pmatrix}$$

The steady-state analysis requires solving:

$$\pi P = \pi \quad \text{with} \quad \sum_i \pi_i = 1$$

bringing:

$$\pi_1 = \frac{1}{1+2(1-p)}, \quad \pi_2 = \pi_{22} = \frac{1-p}{1+2(1-p)}$$

π_1 gives the percentage of time where the first station has got the token, while $1 - \pi_1 = \pi_2 + \pi_{22}$ is the percentage of time where the second station has the token.

The first station has got the token double than the first one when p satisfies:

$$\pi_1 = 2(\pi_2 + \pi_{22})$$

Solving brings: $p = \frac{3}{4}$.

The two stations have the token with equal probability when the chain spends the same amount of time in the first state and the other two states:

$$\pi_1 = \pi_2 + \pi_{22}$$

This is achieved when: $p = \frac{1}{2}$.

Problem 7

On a given communications link, it has been observed that the type of the next packet depends on the previous three packets, such that, if the previous three packets were of voice type, then the next packet is of voice type with probability 0.5. However, if the previous three packets were of data type, then the next packet is also of data type with probability 0.9. In all other cases, the next packet will be of the same type as the previous one with probability 0.6. Use a DTMC to model this problem.

Solution

We need eight states to model all possible situations in this problem. The states will be labeled as $X_3 X_2 X_1$ where the $X_i \in \{V, D\}$. The X_1 refers to the type of the previous packet, X_2 denotes the type of the packet that arrived two slots ago, and X_3 will denote the packet arrival three slots ago.

For instance, state VVV denotes three consecutive Voice packet arrivals in the previous three slots. Hence, the next state after VVV can either be VVV if the next packet arrival is of Voice type, or VVD in case of a Data packet arrival Data. From VVD, the next states can either be: VDV or VDD, and so on. Following this strategy, the solution to this problem is the state-transition diagram of the next figure.

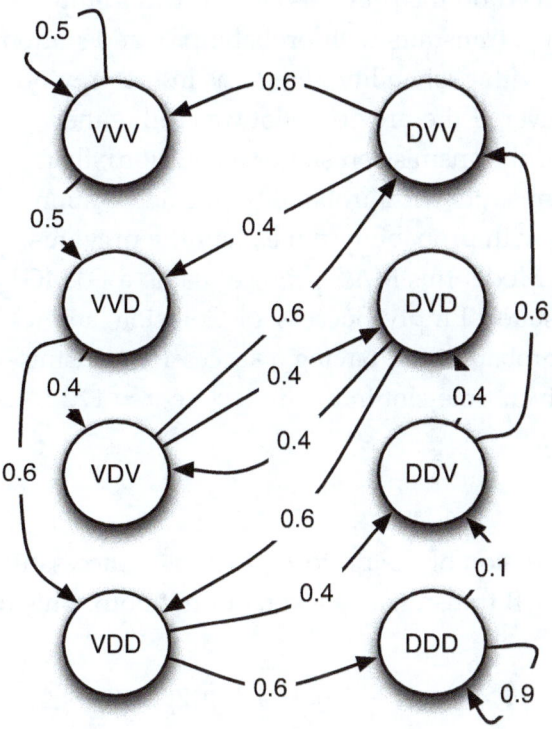

Problem 8

Two end stations A and B share a transmission medium using slotted Aloha. In this medium, packet loss only occurs due to collisions, that is, if the two stations transmit a packet in the same time slot. The stations are assumed to always have packets pending for transmission. In the first version of the MAC protocol, station A transmits a packet with probability p whereas station B transmits a packet with probability $2p$. (a) Derive the probability of having a successful transmission slot, and obtain the value of p that maximizes this probability.

In a second version of the previous MAC protocol, the transmission probability on a time slot depends on what happened on the previous slot: If it was empty, then station A transmits with probability p and station B transmits with probability $2p$ (i.e., as in the previous case). However, if the previous slot was full (either as a successful transmission slot or a collision), then station A transmits with probability $p/2$ and station B transmits with probability p (i.e., half the previous values). (b) Model this MAC protocol using a DTMC with three states: Empty, Success, or Collision; and (c) obtain the probability of having a successful transmission on a given time-slot in the limit. Use $p = 1/2$.

Solution

In the first version of the protocol, there is a successful time slot if either A or B transmits a packet, but not both. This occurs with probability:

$$P_{succ} = p(1 - 2p) + (1 - p)2p = 3p - 4p^2$$

The value of p that maximizes this probability is:

$$\frac{dP_{succ}}{dp} = 0 = 3 - 8p \quad \Rightarrow \quad p = \frac{3}{8}$$

In such a case, the success probability is:

$$P_{succ} = 3\frac{3}{8} - 4\left(\frac{3}{8}\right)^2 = 0.5625$$

In the second version of the protocol, the transmission probabilities for A and B change depending on the result of the previous time slot. Hence, we need to take into account the previous time slot to model what happens in the next one, for instance using a DTMC: We shall use three states to model all possible situations in the previous slot: Empty (if A and B did not transmit a packet), Success (if either A or B transmitted a packet, but not both), and Collision (if both A and B transmitted a packet). The full state-transition diagram is shown in the next figure.

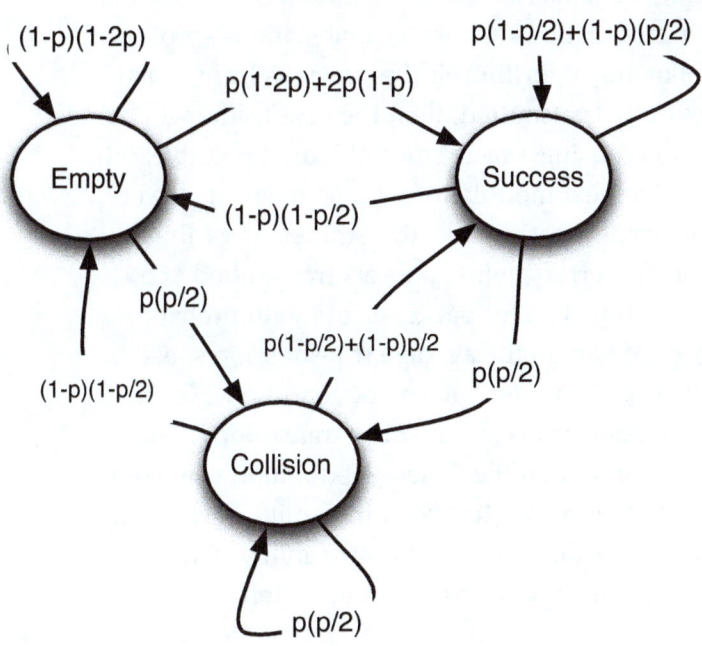

When $p = 1/2$, the transition probability matrix P reduces to:

$$P = \begin{pmatrix} 0 & \frac{1}{2} & \frac{1}{2} \\ \frac{3}{8} & \frac{1}{2} & \frac{1}{8} \\ \frac{3}{8} & \frac{1}{2} & \frac{1}{8} \end{pmatrix}$$

Solving the limiting state probabilities require finding vector π such that:

$$\pi P = \pi \quad \text{with} \quad \sum_i \pi_i = 1$$

which brings:

$$\pi_{empty} = 0.27, \quad \pi_{success} = 0.5, \quad \pi_{collision} = 0.23$$

Problem 9

Consider a wireless network with four possible modulation formats, with different bandwidth capacities: C=1, 2, 4 or 8 Mbit/s. It is observed that modulations with higher capacity values are more error-prone than slower modulations. Consider that stations employ the following algorithm: if the previous frame was successfully transmitted, then the next frame will be transmitted using a faster modulation if possible; otherwise, the next modulation will be the next slower one. Consider that packets transmitted at 1 Mbit/s never suffer errors, while packets transmitted at 2, 4, or 8 4 Mbit/s suffer packet errors with probability 1/4. (a) Obtain the average transmission speed of this protocol under such error conditions. (b) If a certain packet was successfully transmitted, find the probability that the fastest modulation was used; (c) if a certain packet has been transmitted using the 4 Mbit/s modulation, find the probability that the previous frame was transmitted at 2 Mbit/s.

Solution

This problem can be modeled using the DTMC of the next figure, where the states refer to the transmission speed used for transmitting the current packet, namely modulations 1, 2, 4, and 8.

The limiting-state probabilities for this problem are obtained after solving:

$$\pi P = \pi, \quad \sum_i \pi_i = 1$$

where the transition matrix P follows:
$$P = \begin{pmatrix} 0 & 1 & 0 & 0 \\ 1/4 & 0 & 3/4 & 0 \\ 0 & 1/4 & 0 & 3/4 \\ 0 & 0 & 3/4 & 1/4 \end{pmatrix}$$
which produces the following set of equations:
$$\frac{1}{4}\pi_2 = \pi_1$$
$$\pi_1 + \frac{1}{4}\pi_4 = \pi_2$$
$$\frac{3}{4}\pi_4 + \frac{1}{4}\pi_8 = \pi_8$$
$$\pi_1 + \pi_2 + \pi_4 + \pi_8 = 1$$

Solving, we obtain:
$$\pi_1 = \frac{1}{29}, \quad \pi_2 = \frac{4}{29}, \quad \pi_4 = \frac{12}{29}, \quad \pi_8 = \frac{12}{29},$$

The average modulation speed used can be obtained as the state's speed weighted by its probability in the long term:
$$E(Speed) = 1\pi_1 + 2\pi_2 + 4\pi_4 + 8\pi_8 = \frac{153}{29} = 5.28 \text{ Mbit/s}$$

However, some of the frames in this protocol are lost. We can find the probability of a successful transmission making use of the total probability theorem:
$$\begin{aligned} P(Succ) &= P(Succ|Mod=1)P(Mod=1) \\ &+ P(Succ|Mod=2)P(Mod=2) \\ &+ P(Succ|Mod=4)P(Mod=4) \\ &+ P(Succ|Mod=8)P(Mod=8) \\ &= 1\pi_1 + 2\pi_2 + 4\pi_4 + 8\pi_8 = \frac{61}{116} = 0.53 \end{aligned}$$

As shown, only 53% of the transmissions are successful for this protocol. Thus, the effective speed is only 53% of the raw speed (around 2.8 Mbit/s effective), since 47% of the packets are lost. Question (b) requires to apply the Bayes' theorem to see what percentage of successful packets (from such 53% successful) were transmitted using modulation 8 Mbit/s:

$$P(Mod = 8|Succ) = \frac{P(Succ|Mod = 8)P(Mod = 8)}{P(Succ)}$$

$$= \frac{1/4 \times 12/29}{61/116} = \frac{12}{61} = 0.197$$

That means, only about 20% of such 53% successful packets were achieved using the fastest modulation.

Question (c) also requires to apply the Bayes' theorem to see the probability that modulation state 4 comes from a transition from modulation state 2:

$$P(X_{n-1} = 2|X_n = 4) = \frac{P(X_n = 4|X_{n-1} = 2)P(X_{n-1} = 2)}{P(X_n = 4)}$$

$$= \frac{p_{24}\pi_2}{\pi_4} = \frac{1}{4}$$

So only 25% of the times where the chain visits state 4 Mbit/s occur coming from state 2 Mbit/s. Looking at the state-transition diagram shows that the other 75% of the times the chain ends up in state 4 coming from state 8 Mbit/s.

5
Continuous-Time Markov Chains

5.1 CTMCs by example

Continuous-Time Markov Chains (CTMCs) are the continuous-time version of Discrete-Time Markov Chains. CTMCs combine DTMCs with the properties of the exponential distribution. In CTMCs, the amount of time that the Markov chain spends in a state is not fixed but variable, more precisely, exponentially distributed. We shall start with a similar example to that of DTMCs.

Consider again the truck driver who conducts his business in three different cities: London, Oxford, and Nottingham. However, this time the information provided for the example is rather different. The next table provides the *average holding time* of each state, i.e., the average time the truck driver spends in a town once he reaches it.

City	Avg. holding time
London	5 hours
Oxford	10 hours
Nott.	2 hours

Table 5.1: Average holding times for the truck driver example

Now, the truck driver spends a *random amount of time*

in each city, not fixed as in the case of DTMCs. As shown, the truck driver spends more time on average in Oxford (10 hours) and less time on average in Nottingham (only 2 hours).

In addition to the average holding times, the CTMC needs to specify the transition probabilities between states, as shown in the next table.

	Lon.	Oxf.	Nott.
Lon.		0.6	0.4
Oxf.	0.1		0.9
Nott.	0.5	0.5	

Table 5.2: Transition probabilities

Such a table gives the probability of jumping from one city to another once the truck driver leaves that city. The first interesting observation is that there is no transition probability from one state to the same state.

The next figure shows a sample path for this particular CTMC, in other words, a realization of the CTMC process.

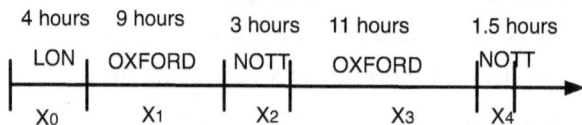

Figure 5.1: Path example for the truck driver example

As observed, the initial state is London ($X_0 = London$), where the truck driver spends exactly 4 hours in that state (remark that the average holding time in London is 5 hours). Next, the chain jumps to state Oxford ($X_1 = Oxford$) where the truck driver spends a total time of 9 hours (note that the average holding time in Oxford is 10 hours). From this state, the chain jumps to state Nott ($X_2 = Nott$) where it stays for 3 hours (in this case, the average holding time is 2 hours). From this state, the chain jumps back to Oxford where it stays for 11 hours (the previous time in Oxford, the truck driver spent 9 hours only). Finally, the chain jumps to the state of Nottingham where it spends 1.5 hours (the last time in Nottingham, the truck driver spent 3 hours). This example shows some properties of this CTMC, namely:

1. The time spent in every state, i.e. the holding time, is a random variable. In CTMCs, this random variable is exponentially distributed with parameter v_i, with $i = 1, 2, 3$; hence the average holding time is $1/v_i$. The parameter v_i refers to the transition rate out of state i. In our example:

    London Avg. holding time: 5 hours $\Rightarrow v_1 = 1/5$ transition/hour

    Oxford Avg. holding time: 10 hours $\Rightarrow v_2 = 1/10$ transition/hour

    Nott. Avg. holding time: 2 hours $\Rightarrow v_3 = 1/2$ transition/hour

2. Second, the next state in the chain depends on the previous one. For instance, when the chain is in the state of Oxford, the next state is Nottingham with probability 0.9, and London with probability 0.1. So, from Oxford, it is more likely to jump to Nottingham than to London.

This information is summarised in the state transition diagram of the next figure. This diagram contains the average holding time of every state once reached, together with the transition probability values from that state to the others.

Clearly, in the long run, the average time spent in every state depends on how reachable this state is from the other states (the transition probabilities), and also the average amount of time spent in a given state once it is reached (average holding time).

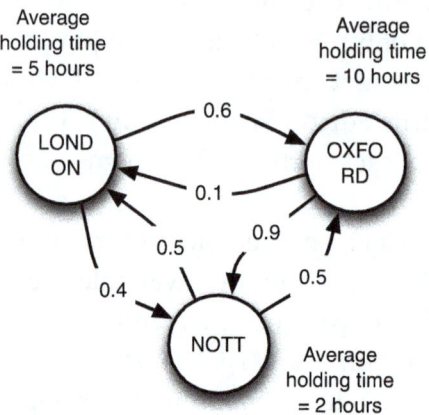

Figure 5.2: State-transition diagram for the truck driver example

Example 1

Consider a CTMC with two states only: A and B. The next table shows a path realization of this Markov chain (time units are hours). Estimate the average holding times of each state and the percentage of time that the chain will spend in every state in the long run.

1.4	2.1	1.6	2.4	0.9	1.5
$X_0 = A$	$X_1 = B$	$X_2 = A$	$X_3 = B$	$X_4 = A$	$X_5 = B$

Solution

This chain alternates from states A and B since it has only two states. The average holding times can be estimated as:

$$\frac{1}{\hat{v}_A} = \frac{1.4 + 1.6 + 0.9}{3} = 1.3 \text{ hours} \Rightarrow \hat{v}_A = 0.77 \text{ trans./hour}$$

$$\frac{1}{\hat{v}_B} = \frac{2.1 + 2.4 + 1.5}{3} = 2 \text{ hours} \Rightarrow \hat{v}_B = 0.5 \text{ trans./hour}$$

We observe that the chain has faster transitions per unit of time from A to B than vice versa, we should expect the chain to spend

CONTINUOUS-TIME MARKOV CHAINS 101

more time in state B than in state A in the long run. In light of this, we can intuitively say that, on average, for every $1.3 + 2 = 3.3$ hour, the chain spends 1.3 hours in state A and 2 hours in state B. Hence, the percentage of time spent in the two states in the long run are:

$$\text{State A:} \quad \frac{1.3}{1.3+2} = 39.4\%$$

$$\text{State B:} \quad \frac{2}{1.3+2} = 60.6\%$$

5.2 The infinitesimal generator

Because the values of $1/v_i$, $i = 1, 2, 3$ represent the average holding time of the i-th state, the value v_i represents the rate at which the chain leaves state i. In the truck driver example, for the state of London, we observe that the average holding time is 5 hours. This means that, once the chain has reached the state of London, it spends an exponentially distributed random time (with a mean of 5 hours) in that state, and then it jumps to another state. This implies that the chain leaves the state of London at a rate of $v_1 = 1/5$ transitions per hour, or 4.8 transitions/day. From the transition probability diagram, we observe that 60% of the times, the truck driver goes to Oxford while the other 40% of the times, his destination is Nottingham. This means that the transition rates towards Oxford and Nottingham are:

$$q_{12} = v_1 p_{12} = 4.8 \times 0.6 = 2.88 \text{ trans/day (London to Oxford)}$$
$$q_{13} = v_1 p_{13} = 4.8 \times 0.4 = 1.92 \text{ trans/day (London to Nott.)}$$

Clearly, $q_{12} + q_{13} = v_i = 4.8$ trans/day (also denoted as $q_{11} = v_1$). Similarly, we can compute the transition rates from Oxford and Nottingham, to the other cities:

$$q_{21} = v_2 p_{21} = 0.24 \text{ trans/day (Oxf. to London)}$$
$$q_{23} = v_2 p_{23} = 2.16 \text{ trans/day (Oxf. to Nott.)}$$

and:

$$q_{31} = v_3 p_{31} = 6 \text{ trans/day (Nott. to London)}$$
$$q_{32} = v_3 p_{32} = 6 \text{ trans/day (Nott. to Oxford)}$$

where $v_2 = 2.4$ transitions per day and $v_3 = 12$ transitions per day. We may collect this information into the following matrix Q:

$$Q = \begin{pmatrix} -4.8 & 2.88 & 1.92 \\ 0.24 & -2.4 & 2.16 \\ 6 & 6 & -12 \end{pmatrix}$$

Matrix Q is often referred to as the *Infinitesimal Generator* or *Intensity Matrix*. As noted, the rows of Q add up to zero.

Example 2

Obtain the infinitesimal generator for the two-state CTMC of Example 1.

Solution

The intensity matrix in this example is:

$$Q = \begin{pmatrix} -0.77 & 0.77 \\ 0.5 & -0.5 \end{pmatrix}$$

The intensities of matrix Q for the truck driver example can also be depicted in the so-called *state-transition-rate diagram* (see next figure).

The information provided from the state-transition-rate diagram is many-fold. For the state of London for example, the diagram reveals that:

1. Once the chain arrives at state London, two jumping events may occur: (1) the chain jumps to state Oxford or (2) Nottingham. These two events are exponentially distributed random variables with parameters 2.88 and 1.92 transition/hour respectively. It is worth remarking from Chapter 2 that the minimum of two exponential

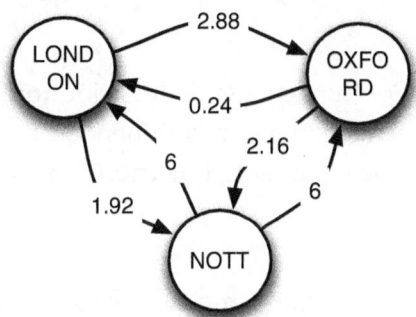

Figure 5.3: State-transition rate diagram for the truck driver example.

random variables $X_1 \sim exp(\lambda_1)$ and $X_2 \sim exp(\lambda_2)$ is also exponentially-distributed with rate $\lambda_m = \lambda_1 + \lambda_2$.

2. The holding time in state London is then characterized as the minimum of these two random variables, i.e. exponentially distributed with parameter $v_1 = 2.88 + 1.92 = 4.8$ transitions per day. In other words:

$$v_1 p_{12} + v_1 p_{13} = v_1$$

3. The next state after London can be obtained as the comparison between such two exponentially-distributed random variables: The next state is Oxford with probability: $\frac{2.88}{2.88+1.92} = 0.6$ or Nottingham with probability: $\frac{1.92}{2.88+1.92} = 0.4$. In other words:

$$\frac{v_1 p_{12}}{v_1 p_{12} + v_1 p_{13}} = p_{12}$$

It is worth remarking from Chapter 2 that, when comparing two exponential random variables $X_1 \sim exp(\lambda_1)$ and $X_2 \sim exp(\lambda_2)$, then $X_1 < X_2$ with probability

$$\frac{\lambda_1}{\lambda_1 + \lambda_2}$$

4. Finally, the average time spent in London is: $\frac{1}{v_1} = 0.2083$ days (or 5 hours)

Thanks to the properties of the exponential distribution, CTMCs are very tractable. The reader is recommended to review the properties of the exponential distribution explained in Chapter 2.

5.3 Definition and Chapman-Kolmogorov equations

Let $X(t)$ denote the state at which the truck driver is at time t. In our example, $X(t)$ may take three possible values: London, Oxford and Nottingham. For simplicity, we shall consider $X(t) = 1$ for London, $X(t) = 2$ for Oxford and $X(t) = 3$ for Nottingham.

Definition The stochastic process $\{X(t), t \geq 0\}$ is a Continuous-Time Markov Chain if for all $t_0, t_1, t_2, \ldots$ and all states recorded $s_0, s_1, s_2, \ldots$, the following property is satisfied:

$$P(X(t_{n+1}) = s_{n+1} | X(t_n) = s_n, \ldots, X(t_0) = s_0) = P(X(t_{n+1}) = s_{n+1} | X(t_n) = s_n) \quad (5.1)$$

The above equation is the continuous-time version of the Markov property studied in DTMCs. The idea is that the future state at time $t + s$ does not depend on the whole past set of states, but only the very last one.

For notation purposes, we shall consider the following transition probability functions:

$p_j(t) = P(X(t) = j)$ which refers to the probability that the chain is in state j at time t. In this case, the $p_j(t)$ must satisfy:

$$\sum_j p_j(t) = 1, \quad 0 \leq p_j(t) \leq 1 \quad (5.2)$$

since the chain must be in any state at all times t. We may collect the set of probabilities $p_j(t)$ into a vector $p(t)$:

$$p(t) = \begin{pmatrix} p_1(t) & p_2(t) & p_3(t) & \ldots \end{pmatrix} \quad (5.3)$$

$p_{ij}(t) = P(X(t+s) = j|X(s) = i)$ which refers to the probability that the Markov chain, whose actual state at time s is i, will be in state j after t units of time. The probability values $p_{ij}(t)$ values must also satisfy:

$$\sum_j p_{ij}(t) = 1, \quad 0 \leq p_{ij}(t) \leq 1 \tag{5.4}$$

which refers to the fact that, after some time t, the chain has either jumped to a different state or remained in the same state. We may also collect the set of probabilities $p_{ij}(t)$ into a matrix $P(t)$:

$$P(t) = \begin{pmatrix} p_{11}(t) & p_{12}(t) & p_{13}(t) & \cdots \\ p_{21}(t) & p_{22}(t) & p_{23}(t) & \cdots \\ p_{31}(t) & p_{32}(t) & p_{33}(t) & \cdots \\ \vdots & \vdots & \vdots & \ddots \end{pmatrix} \tag{5.5}$$

We note that this matrix $P(t)$ is different from the infinitesimal generator Q.

It can be further demonstrated that:

$$P(s+t) = P(s)P(t) \tag{5.6}$$

which is the CTMC-version of the *Chapman-Kolmogorov equations* in a matrix form. Alternatively, the Chapman-Kolmogorov equations may be formulated as:

$$p_{ij}(s+t) = \sum_k p_{ik}(s) p_{kj}(t) \tag{5.7}$$

for all possible states i, j, k.

The next section shows how to solve $p(t)$ for all values of t in the so-called transient behavior of the CTMC. To do so, we will make use of the Infinitesimal Generator Q introduced before.

5.4 The transient behavior of a CTMC

Consider a very small amount of time Δt, and let $p_i(t + \Delta t)$ denote the probability of being in state i at time $t + \Delta t$.

Then:

$$p_i(t+\Delta t) = p_i(t)\left(1 - \sum_{j \neq i} v_i p_{ij} \Delta t\right) + \sum_{j \neq i} p_j(t) v_j p_{ji} \Delta t, \quad i = 0, 1, 2, \ldots \tag{5.8}$$

This equation is explained as follows: The chain ends up in state i at time $t + \Delta t$ if it was previously in state i at time t and did not make a transition during Δt to any other state (this is the first term in the equation), or if it was in any other state j at time t and made a transition from that state j into state i within Δt.

Rearranging the terms in eq. 5.8, we obtain:

$$\frac{p_i(t+\Delta t) - p_i(t)}{\Delta t} = -p_i(t) v_i \sum_{j \neq i} p_{ij} + \sum_{j \neq i} p_j(t) v_j p_{ji}, \quad i = 0, 1, 2, \ldots \tag{5.9}$$

Taking the limit when $\Delta t \to 0$:

$$\frac{dp_i(t)}{dt} = -p_i(t) v_i \sum_{j \neq i} p_{ij} + \sum_{j \neq i} p_j(t) v_j p_{ji}, \quad i = 0, 1, 2, \ldots \tag{5.10}$$

Now, we can make use of the previously defined vectors:

$$p(t) = \begin{pmatrix} p_1(t) & p_2(t) & p_3(t) & \cdots \end{pmatrix} \tag{5.11}$$

$$\frac{dp(t)}{dt} = \begin{pmatrix} \frac{dp_1(t)}{dt} & \frac{dp_2(t)}{dt} & \frac{dp_3(t)}{dt} & \cdots \end{pmatrix} \tag{5.12}$$

and

$$Q = \begin{pmatrix} -q_{11} & q_{12} & q_{13} & \cdots \\ q_{21} & -q_{22} & q_{23} & \cdots \\ q_{31} & q_{32} & -q_{33} & \cdots \\ \vdots & \vdots & \vdots & \ddots \end{pmatrix} \tag{5.13}$$

which is the infinitesimal generator Q defined before (remark that the q_{ij} values satisfy: $q_{ij} = v_i p_{ij}$). Then, the set of equations 5.10 in matrix form becomes:

$$\frac{dp(t)}{dt} = p(t) Q \tag{5.14}$$

Solving this differential equation brings:

$$p(t) = p(0) e^{Qt} \tag{5.15}$$

where $p(0)$ is the initial distribution vector and the exponential of a matrix e^{Qt} is computed as:

$$e^{Qt} = I + \sum_{k=1}^{\infty} \frac{Q^k t^k}{k!} \qquad (5.16)$$

Example 3

In the truck driver example, obtain the probability vector $p(t)$ that the truck driver stays in London at times: $t = 0.01, 0.1, 0.4, 1, 10, 100$ days. Consider that the truck driver starts from the state of London at time $t = 0$.

Solution

This requires to compute eq. 5.15 for different values of t:

$$p(t) = p(0)e^{Qt}$$

Here:

$$p(0) = \begin{pmatrix} 1 & 0 & 0 \end{pmatrix}$$

and the infinitesimal generator Q is:

$$Q = \begin{pmatrix} -4.8 & 2.88 & 1.92 \\ 0.24 & -2.4 & 2.16 \\ 6 & 6 & -12 \end{pmatrix}$$

We may compute eq. 5.15:

$$p(t = 0.01) = p(0)e^{Q \cdot 0.01} = \begin{pmatrix} 0.95 & 0.03 & 0.02 \end{pmatrix}$$

Since the average holding time in London is 5 hours (or 0.2 days), the chain is very likely to remain in the state of London (with 95% probability) after 0.01 days. It is worth remarking that the time units must be consistent, that is, if t is in days, then the rates q_{ij} must be specified in transitions per day.

At the time $t = 0.1$, we have a much larger probability of having left London:

$$p(t = 0.1) = p(0)e^{Q \cdot 0.1} = \begin{pmatrix} 0.65 & 0.24 & 0.11 \end{pmatrix}$$

At time $t = 0.4$, we have that the truck driver will most likely be in Oxford:

$$p(t = 0.4) = p(0)e^{Q \cdot 0.4} = \begin{pmatrix} 0.30 & 0.55 & 0.15 \end{pmatrix}$$

At time $t = 1$, we have the following probability values:

$$p(t = 1) = p(0)e^{Q \cdot 1} = \begin{pmatrix} 0.2216 & 0.6296 & 0.1487 \end{pmatrix}$$

At time $t = 10$ and $t = 100$ days, we obtain:

$$p(t = 10) \approx p(t = 100) = \begin{pmatrix} 0.2177 & 0.6334 & 0.1489 \end{pmatrix}$$

Essentially, after a sufficiently large amount of time, the initial distribution vector is not important and the probability values converge to the so-called steady-state probabilities, in the same way as in DTMCs.

The next section is devoted to the steady-state analysis of CTMCs.

5.5 Steady-state analysis and balance equations

In the steady state, $p_j(t) \to p_j$ and:

$$\lim_{t \to \infty} \frac{dp_i(t)}{dt} = 0 \qquad (5.17)$$

which yields:

$$pQ = 0, \quad \text{with} \quad \sum_i p_i = 1 \qquad (5.18)$$

where p is the *steady-state vector* of the CTMC.

It is worth noticing that this result is very similar to the steady-state analysis of DTMCs, where the steady-state vector π is obtained from:

$$\pi P = \pi, \quad \text{with} \quad \sum_i \pi_i = 1$$

Example 4

Obtain the steady-state vector p in the truck driver example.

Solution

Solving eq. 5.18:

$$\begin{pmatrix} p_1 & p_2 & p_3 \end{pmatrix} \begin{pmatrix} -4.8 & 2.88 & 1.92 \\ 0.24 & -2.4 & 2.16 \\ 6 & 6 & -12 \end{pmatrix} = \begin{pmatrix} 0 & 0 & 0 \end{pmatrix}$$

produces the following set of equations:

$$\begin{aligned} -4.8p_1 + 0.24p_2 + 6p_3 &= 0 \\ 2.88p_1 - 2.4p_2 + 6p_3 &= 0 \\ 1.92p_1 + 2.16p_2 - 12p_3 &= 0 \\ p_1 + p_2 + p_3 &= 1 \end{aligned}$$

The three first questions are linearly dependent (as in DTMCs) so we may remove one of them (for instance the third one). Then, we have a set of three equations with three unknowns. After solving them, we obtain:

$$p_1 = 0.2177 \quad p_2 = 0.6334 \quad p_3 = 0.1489$$

This result was also obtained in the previous section, after computing the transient analysis for a sufficiently large amount of time (in the example $t = 10$ or $t = 100$).

The steady-state set of equations given by Eq. 5.18 can also be formulated as:

$$\begin{aligned} 0 &= -p_i q_{ii} + \sum_{j \neq i} p_j q_{ji}, \quad i = 0, 1, 2, \ldots \\ 1 &= \sum_i p_i \end{aligned} \quad (5.19)$$

Here, we have assumed that the limiting-state probabilities p_i, $i = 0, 1, 2, \ldots$ indeed exist. However, the limiting-state probabilities do not always exist. A sufficient condition for that requires (similarly to that specified in DTMCs):

1. All states of the Markov chain intercommunicate. In other words, for all states i and j of the chain, there is some positive probability of ever visiting state j starting from state i.
2. The Markov chain is positive recurrent in the sense that, starting in any state i, the average time to return to that state is finite.

If these two conditions are met, then the limiting-state probabilities will exist and satisfy eqs. 5.19. In addition, such probabilities also provide the portion of time that the process stays in every state in the long run. In addition, eqs. 5.19 are also known as *the balance equations*.

> ### Example 5
>
> Apply the balance equations to obtain the limiting state probabilities of the following CTMC:
>
>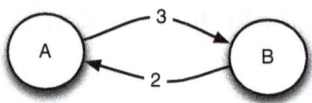

> ### Solution
>
> The balance equations state that the flux out of a state must equal the flux into that state. This comprises the following:
>
> $$3p_A = 2p_B$$
>
> This equation, together with:
>
> $$p_A + p_B = 1$$
>
> comprises a set of two equations with two unknowns whose solution gives the limiting-state probabilities p_A and p_B. Solving, we obtain:
>
> $$p_A = \frac{2}{5}, \quad p_B = \frac{3}{5}$$

Finally, it is worth noticing that this solution can also be found from Eq. 5.18:

$$\begin{pmatrix} p_A & p_B \end{pmatrix} \begin{pmatrix} -3 & 3 \\ 2 & -2 \end{pmatrix} = \begin{pmatrix} 0 & 0 \end{pmatrix}$$

which produces the same set of equations:

$$-3p_A + 2p_B = 0$$
$$3p_A - 2p_B = 0$$
$$p_A + p_B = 1$$

where the first two are linearly dependent.

5.6 First-passage times

We define the average first passage time m_{ij} from state i to state j as the average time required to visit state j assuming that the Markov chain is currently in state i. Using conditioning, it can be proved that finding this value requires to solve the following set of equations:

$$m_{ij} = \frac{1}{q_{ii}} + \sum_{k \neq j} \frac{q_{ik}}{q_{ii}} m_{kj}, \quad i = 0, 1, 2, \ldots \quad (5.20)$$

This equation states that the average first passage time from state i to state j requires to first take into account the average holding time spent on state i (this is the first term in the equation: $1/q_{ii}$), and then compute the average first passage time m_{kj} for all possible intermediate states k weighted by the probability of jumping from i to that state k (that weight is given by the probability of jumping to state k before any other state q_{ik}/q_{ii}).

Example 6

Obtain the average first passage time from London to Nottingham in the truck driver example, i.e. m_{13}.

Solution

To solve this example, we need to apply Eq. 5.20:

$$m_{13} = \frac{1}{q_{11}} + \frac{q_{12}}{q_{11}} m_{23} = \frac{1}{4.8} + \frac{2.88}{4.8} m_{23}$$

where:

$$m_{23} = \frac{1}{2.4} + \frac{0.24}{2.4} m_{13}$$

Thus:

$$m_{13} = \frac{1}{4.8} + \frac{2.88}{4.8}\left(\frac{1}{2.4} + \frac{0.24}{2.4} m_{13}\right)$$

Hence:

$$m_{13} = \frac{\frac{1}{4.8} + \frac{2.88}{4.8} \times \frac{0.24}{2.4}}{1 - \frac{2.88}{4.8} \times \frac{0.24}{2.4}} = 0.4876 \text{ day}$$

and

$$m_{23} = \frac{1}{2.4} + \frac{0.24}{2.4} \times 0.4876 = 0.4654 \text{ day}$$

We can further rearrange the set of two equations with two unknowns:

$$\begin{aligned} -4.8 m_{13} + 2.88 m_{23} &= -1 \\ 0.24 m_{13} - 2.4 m_{23} &= -1 \end{aligned}$$

(5.21)

which is equivalent to:

$$\begin{pmatrix} -4.8 & 2.88 \\ 0.24 & -2.4 \end{pmatrix} \begin{pmatrix} m_{13} \\ m_{23} \end{pmatrix} = \begin{pmatrix} -1 \\ -1 \end{pmatrix}$$

It is worth noticing that the first matrix is a subset of the infinitesimal generator Q, where the third row and column have been removed.

Essentially, to find the average first passage times from all states to some destination state j, we just need to construct a matrix A_{fpt} which is basically matrix Q without the j-th row and column, and solve:

$$A_{fpt} m^T = -1^T$$

where m is the row-vector:

$$m = \begin{pmatrix} m_{1j} & m_{2j} & \cdots & m_{(j-1)j} & m_{(j+1)j} & \cdots & m_{Nj} \end{pmatrix} \quad (5.22)$$

and 1 is the row-vector:

$$1 = \begin{pmatrix} 1 & 1 & \cdots & 1 \end{pmatrix} \quad (5.23)$$

5.7 Further problems

Problem 1

The next figure shows the state-transition-rate diagram of a given CTMC. Here, the numbers represent the transition rates from one state to another in events per second. At a given time t, the chain is known to be in state A. Answer the following questions: (a) What is the probability that the chain spends more than 0.1 seconds in that state? (b) For how long is the chain going to be on state A on average? (c) What is the probability that the next state is B? What about state C? (d) Can you find the average first passage time from state C to state A? (e) Can you find the steady-state probabilities of states A, B, and C?

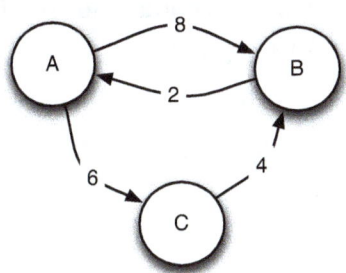

Solution

In state A, the chain may jump to state B at a rate of 8 jump/sec or to state C at a rate of 6 jump/sec. Hence, the total outgoing rate is $v_A = 6 + 8 = 14$ jump/sec. This means that the chain spends an exponentially distributed random time in state A before jumping to any other state (B or C). Let H_A refer to such holding time random variable. Then:

$$P(H_A > 0.1) = e^{-v_A t} = e^{-14 \times 0.1} = 0.25$$

and the average holding time in state A is:

$$E(H_A) = \frac{1}{v_A} = \frac{1}{14} \text{ secs} = 0.0714 \text{ secs}$$

From state A, the next state is B with probability:

$$\frac{q_{AB}}{\sum_k q_{Ak}} = \frac{8}{6+8} = 0.57$$

and C with probability:

$$\frac{q_{AC}}{\sum_k q_{Ak}} = \frac{6}{6+8} = 0.43$$

The average first-passage time from state C to state A can be easily obtained from the observation that, once in state C, the chain must jump to state B, and from B to A mandatory. Hence, the first-passage time from C to A equals the sum of the average holding time in states C and A:

$$m_{CA} = \frac{1}{v_C} + \frac{1}{v_B} = \frac{1}{4} + \frac{1}{2} = 0.75 \text{ secs}$$

Finally, the steady-state probabilities come after solving the following set of equations (balance equations):

$$14 p_A = 2 p_B$$
$$6 p_A = 4 p_C$$
$$p_A + p_B + p_C = 1$$

After solving, we obtain:

$$p_A = \frac{2}{19}, \quad p_B = \frac{14}{19}, \quad p_C = \frac{3}{19}$$

Problem 2

In a computer lab, three servers are always online. Each server is observed to break down once every week on average. There is a single technician who has instructions to bring them all online when all three computers are down. This technician takes an exponentially distributed random time with a mean of two days to bring the three servers back online. (a) Use a CTMC to model this problem and write down the infinitesimal generator Q; (b) solve the limiting-state probabilities to see what percentage of time all three servers are down in the long run; (c) find also the average number of servers down, and the average time to have all three servers down once the technician has repaired the three servers.

Solution

To build a CTMC model, we first need to identify how many different states the system may have and the transitions between them. Essentially, four states are needed, which identify all possible failure situations: Zero, one, two, or three servers down. The next stage is to identify the transition rates between states. Essentially, these depend on how many servers are online. For instance, in the state Zero, we have all three servers online, so all of them may fail. This means that there is a transition to state One (one server down) as soon as any of the three servers break down. Since the server failure times are exponentially distributed with the rate $\lambda = \frac{1}{7}$ failure/day, then the minimum of such three random variables is also exponentially distributed with rate 3λ. The transition from state One (server down) to state Two (servers down) is also exponentially distributed with a rate of 2λ (the minimum of two exponential random variables with a rate of λ each). From state Two to state Three, the transition rate is only λ (one server alive only). Finally, the transition from state Three back to state Zero depends on how fast the technician repairs all three machines. This is modeled with an exponential random variable with rate $\mu = \frac{1}{2}$ repair/day. The next figure shows the state-transition-rate diagram of this CTMC.

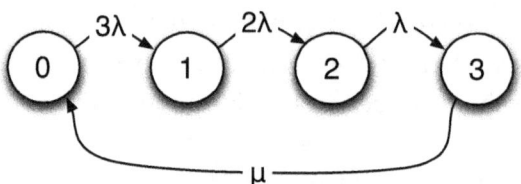

The infinitesimal generator, i.e. matrix Q is:

$$Q = \begin{pmatrix} -3\lambda & 3\lambda & 0 & 0 \\ 0 & -2\lambda & 2\lambda & 0 \\ 0 & 0 & -\lambda & \lambda \\ \mu & 0 & 0 & -\mu \end{pmatrix}$$

The balance equations for this problem are:

$$\begin{aligned} 3\lambda p_0 &= 2\lambda p_1 & \Rightarrow \quad p_1 &= \frac{3}{2} p_0 \\ 2\lambda p_1 &= \lambda p_2 & \Rightarrow \quad p_2 &= 2 p_1 = 3 p_0 \\ \lambda p_2 &= \mu p_3 & \Rightarrow \quad p_3 &= \frac{\lambda}{\mu} p_2 = 3 \frac{\lambda}{\mu} p_0 \\ \mu p_3 &= 3\lambda p_0 & \Rightarrow \quad p_3 &= 3 \frac{\lambda}{\mu} p_0 \end{aligned}$$

(5.24)

Remark that the balance equations state that the flux of transitions out of a state must equal the flux of transitions into that same state in the long run.

We only need the first three equations of the previous four, together with:

$$p_0 + p_1 + p_2 + p_3 = 1$$

Solving for p_0:

$$p_0 \left(1 + \frac{3}{2} + 3 + \frac{3\lambda}{\mu} \right) = 1$$

which brings:

$$p_0 = \frac{1}{\frac{11}{2} + \frac{3\lambda}{\mu}} = \frac{2\mu}{11\mu + 6\lambda} = \frac{2\frac{1}{2}}{11\frac{1}{2} + 6\frac{1}{7}} = \frac{14}{89}$$

Solving for p_1, p_2 and p_3:

$$p_0 = \frac{14}{89}, \quad p_1 = \frac{3}{2}p_0 = \frac{21}{89}, \quad p_2 = 3p_0 = \frac{42}{89}$$

$$p_3 = 3\frac{\lambda}{\mu}p_0 = \frac{12}{89}$$

In the long run, the percentage of time when all three servers are down is:

$$p_3 = 0.135 \quad (13.5\%)$$

and the average number of servers down is:

$$E(N) = 0p_0 + 1p_1 + 2p_2 + 3p_3 = \frac{21 + 2 \times 42 + 3 \times 12}{89}$$
$$= \frac{141}{89} = 1.58 \text{ servers down}$$

Finally, the average first passage time from state 0 to state 3 can be obtained using Eq. 5.20. Alternatively, we can note that, once the technician has repaired all three servers, the process spends an average amount of time of $\frac{1}{3\lambda}$ on state 0, then spends an average of time of $\frac{1}{2\lambda}$ on state 1 and an average amount of time of $\frac{1}{\lambda}$ on state 2. After this time, the process arrives at state 3 (all servers down). This total time accounts for:

$$\frac{1}{3\lambda} + \frac{1}{2\lambda} + \frac{1}{\lambda} = \frac{11}{6\lambda} = 12.83 \text{ days}$$

After this, the technician spends an average amount of time of 2 days to repair the three machines altogether. Hence, the total amount of time when all machines are down is:

$$\frac{2}{2 + 12.83} = 0.135$$

which is the same amount as p_3.

Problem 3

Consider the same three servers as before. This time there are two technicians rather than one, and these technicians decide to repair the servers as soon as they break down. However, we consider that a technician takes 2 days on average to repair one server (in the previous example, a single technician could repair 3 servers in 2 days on average, i.e. much faster). Using the same assumptions regarding the inter-failure events, (a) formulate the new CTMC and matrix Q, and (b) solve the limiting state probabilities and find the average number of servers down in the long run and the percentage of time where all servers are down.

Solution

Again, the number of states (or system situations) is four, ranging from state Zero (no server down) to Three (all servers down). The difference with the previous problem lies in the transitions between the states, as shown in the next figure.

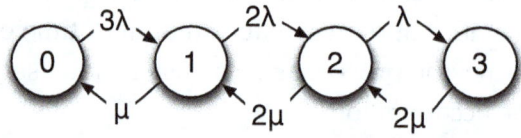

Essentially, the forward transition rates (from state i to state $i + 1$), remain the same as before. However, we have a new transition from state 1 to state 0, which occurs at rate μ. This transition rate accounts for the rate at which a single technician repairs the server down (only one technician is repairing the broken server, the other technician is idle). If a new server breaks down before the first broken server is repaired (this means that a transition from state 1 to state 2 occurs before a transition from 1 to 0), then the second technician comes into action. Hence, the transition rate from state 2 to state 1 is 2μ which refers to the minimum between two exponential random variables with rate μ each (either the first or the second technician brings one server up again). Finally, in state 3, the transition rate back to state 2 occurs at the rate 2μ, the same as before. Essentially, in state 3 we have three broken servers but the two technicians can only handle one server each, thus the third broken server must wait until any of the other two servers is brought back online and a technician is freed to begin repairing it. It is further worth noting that, in state 1 one technician is in the process of repairing the broken machine. If a transition to state 2 occurs, a second technician starts repairing the second broken machine. The interesting thing about the memoryless property of the exponential distribution is that the first technician has already spent some time repairing the first machine when the second technician begins repairing his machine. However, such history is not important and the system renews itself upon arrival in state 2. In other words, when the chain reaches state 2, the transition rate back to state 1 is $\mu + \mu = 2\mu$ regardless of the amount of time that the first technician has spent on repairing the first machine during previous state 1. It is like if the first technician starts his job again on the first machine. This is only possible thanks to the memoryless property of the exponential distribution and is a clear advantage in analyzing CTMCs.

The new infinitesimal generator, i.e. matrix Q is:

$$Q = \begin{pmatrix} -3\lambda & 3\lambda & 0 & 0 \\ \mu & -(2\lambda+\mu) & 2\lambda & 0 \\ 0 & 2\mu & -(\lambda+2\mu) & \lambda \\ 0 & 0 & 2\mu & -2\mu \end{pmatrix}$$

We note that the rows in Q must add up to unity.
The new balance equations are then:

$$3\lambda p_0 = \mu p_1 \Rightarrow p_1 = 3\frac{\lambda}{\mu}p_0$$
$$(2\lambda+\mu)p_1 = 3\lambda p_0 + 2\mu p_2$$
$$(\lambda+2\mu)p_2 = 2\lambda p_1 + 2\mu p_3$$

The second equation can be reduced to:

$$2\lambda p_1 = 2\mu p_2 \quad \Rightarrow \quad p_2 = \frac{\lambda}{\mu} p_1 = 3\left(\frac{\lambda}{\mu}\right)^2 p_0$$

since the first equation states that $3\lambda p_0 = \mu p_1$, hence these items can be removed from both sides of the equation. The same reasoning applies to the third equation, leading to:

$$\lambda p_2 = 2\mu p_3 \quad \Rightarrow \quad p_3 = \frac{\lambda}{2\mu} p_2 = \frac{3}{2}\left(\frac{\lambda}{\mu}\right)^3 p_0$$

Now, we can solve for p_0 using $p_0 + p_1 + p_2 + p_3 = 1$, which yields:

$$p_0 = \frac{1}{1 + 3\frac{\lambda}{\mu} + 3\left(\frac{\lambda}{\mu}\right)^2 + \frac{3}{2}\left(\frac{\lambda}{\mu}\right)^3}$$

where:

$$\frac{\lambda}{\mu} = \frac{1/7}{1/2} = \frac{2}{7}$$

After some calculus, we obtain:

$$p_0 = \frac{343}{733}, \quad p_1 = \frac{294}{733}, \quad p_2 = \frac{84}{733}, \quad p_3 = \frac{12}{733}$$

Thus, the percentage of time when all three servers are down is now $p_3 = 0.016$; and the average number of servers down over time is:

$$E(N) = 0p_0 + 1p_1 + 2p_2 + 3p_3 = \frac{498}{733} = 0.68 \text{ servers}$$

Problem 4

Consider a scenario with two different servers: Machine i, $i = 1, 2$ operates for an exponential time with rate $\lambda_i = i \cdot \lambda$ (where $\lambda = 2$ failure/year) and then fails; its repair time is exponential with rate $\mu_i = (3 - i)\mu$ (where $\mu = 1$ repair/month). The two servers behave independently. Define a Markov chain that describes the system, and solve the limiting-state probabilities.

Solution

In this case, we cannot use three states that refer to zero, one, or two servers down since the two servers are different. We need to consider two separate states for the case of one server down. Hence, we define four states: "no server down", "server $i = 1$ down", "server $i = 2$ down", "both servers down". The transition rates are also different from the state "no server down" to "one server down", depending on which server breaks down. The next figure shows the state-transition-rate diagram for this scenario.

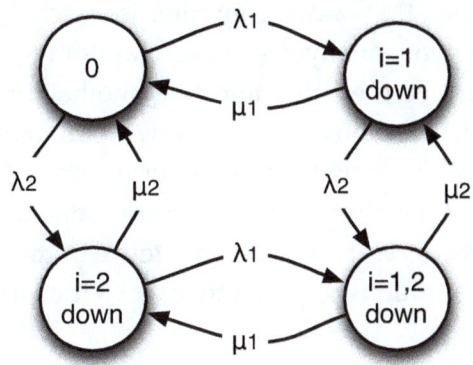

The balance equations for this CTMC are:

$$(\lambda_1 + \lambda_2)p_0 = \mu_1 p_1 + \mu_2 p_2$$
$$(\lambda_2 + \mu_1)p_1 = \lambda_1 p_0 + \mu_2 p_{12}$$
$$(\mu_1 + \mu_2)p_{12} = \lambda_2 p_1 + \lambda_1 p_2$$
$$p_0 + p_1 + p_2 + p_{12} = 1$$

which can be reduced to:

$$6p_0 = 24p_1 + 12p_2$$
$$28p_1 = 2p_0 + 24p_{12}$$
$$36p_{12} = 4p_1 + 2p_2$$
$$p_0 + p_1 + p_2 + p_{12} = 1$$

since $\lambda_1 = \lambda = 2$ failure/year, $\lambda_2 = 2\lambda = 4$ failure/year, $\mu_1 = (3 -$

1) $\mu = 2\mu = 2$ repair/month $= 24$ repair/year and $\mu_2 = (3-2)\mu = 1$ repair/month $= 12$ repair/year.
Solving the balance equations brings:

$$p_1 = 0.718, \quad p_2 = 0.059, \quad p_2 = 0.205, \quad p_{12} = 0.018$$

Problem 5

Consider a collection of computers infected by a virus in the Internet. Each computer can either infect another computer or break down independently from others. We assume that once a computer is infected, it waits for an exponentially distributed random time with parameter λ before either infecting another one (with probability p) or breaking down (with probability $1 - p$). Let $X(t)$ denote the number of computers infected at time t. Draw the state-transition-rate diagram of this process and write down a few comments regarding the conditions required to remove the virus from the Internet.

Solution

In this case, the states characterize the number of infected computers in the Internet, thus ranging from $n = 0, 1, 2, \ldots, N$, where N is the total number of computers in the Internet. The next figure shows the state-transition-rate diagram for this CTMC.

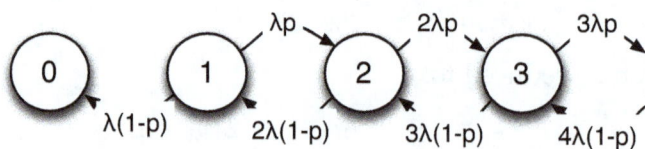

As shown, both the births (i.e. from state n to state $n + 1$, new computer infected) and the deaths (i.e. from state n to state $n - 1$, infected computer down) depend on the current state, that is, on the actual number of infected computers at that time. In other words:

$$q_{n,n+1} = n\lambda p, \quad q_{n,n-1} = n\lambda(1-p), \quad n = 1,\ldots, N-1$$

In this CTMC, the death rates must be greater than the birth rates to force the chain to go to state o (none infected) before reaching state N (all infected). This means $p < 0.5$. The The amount of time required to remove that virus from the Internet depends on both the initial state, that is, the number of infected computers at time $t = 0$ and also on the value of p which determines how fast deaths occur concerning births. Essentially, a death occurs before a birth with probability:

$$\frac{n\lambda(1-p)}{n\lambda p + n\lambda(1-p)} = 1 - p$$

Problem 6

The hardware of a given computer has two resources: CPU and GPU. The type of jobs attended by the operating system requires the CPU first and then the GPU. Both CPU and GPU can hold only one job at a time. When a job is completed in the CPU, then it goes to the GPU if this one is empty; otherwise, it must wait in the CPU until the GPU is free (see figure below).

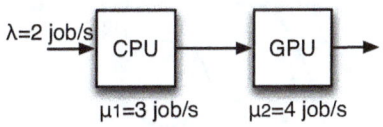

Consider that the operating system offers jobs to the CPU according to a Poisson process with rate $\lambda = 2$ job/s. An arriving job is accepted only if there is no other item in the CPU, otherwise it is lost from the system. The service times spent by jobs in the CPU and GPU are exponentially distributed with parameters $\mu_1 = 3$ job/s and $\mu_2 = 4$ job/s respectively. (a)

Draw the state-transition-rate diagram of the process;
(b) find matrix Q and the limiting-state probabilities;
(c) determine the percentage of jobs rejected because the CPU is busy, and also the rate at which jobs enter the CPU.

Hint: Label each state as (m, n), where m is the number of items in the CPU and n is the number of items in the GPU.

Solution

We need the following states to characterize the different situations of this problem: State 00 refers to no job in the system; state 10 refers to a job in the CPU, while the GPU is empty; state 01 refers to a job in the GPU and no job in the CPU; state 11 refers to a job in both CPU and GPU. Finally, another state is needed to model the situation where both CPU and GPU have one job each, but the job in the CPU is already completed and must wait until the job in the GPU finishes; we shall call this state F1, F for "finished". The complete CTMC model is shown below.

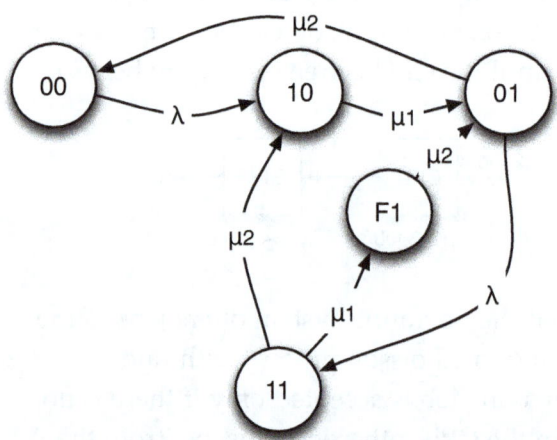

The infinitesimal generator Q for this problem is (the row and column order is 00, 10, 01, 11, and F1):

$$Q = \begin{pmatrix} -\lambda & \lambda & 0 & 0 & 0 \\ 0 & -\mu_1 & \mu_1 & 0 & 0 \\ \mu_2 & 0 & -(\lambda+\mu_2) & \lambda & 0 \\ 0 & \mu_2 & 0 & -(\mu_1+\mu_2) & \mu_1 \\ 0 & 0 & \mu_2 & 0 & -\mu_2 \end{pmatrix}$$

Solving the steady-state, we find:

$$p_{00} = 0.3836, \quad p_{10} = 0.3288, \quad p_{01} = 0.1918$$

$$p_{11} = 0.0548, \quad p_{F1} = 0.0411$$

If a new job arrives and the process is, at that particular moment, in any of 10, 11 or F1, then the job is rejected since the CPU is busy. This occurs with probability:

$$p_{reject} = p_{10} + p_{11} + p_{F1} = 0.4247$$

Hence, the rate λ_{accept} at which jobs are accepted in the CPU is:

$$\lambda_{accept} = \lambda(1 - p_{reject}) = 0.8494 \text{ job/s}$$

Essentially, job arrivals occur following a Poisson process with rate λ (inter-arrival times are exponentially distributed with mean $1lambda$). Thanks to the *Poisson Arrivals See Time Averages* (or PASTA property), Poisson arrivals observe the system busy with probability p_{reject}. Therefore, those job arrivals that observe the system busy are rejected. On average, the net input rate is $\lambda(1 - p_{reject})$.

6
Classical queueing theory

6.1 Introduction by example

John is a Ph.D student enrolled in the Computer Science department. At lunch time, most Ph.D. students bring their own food and have lunch together in a common meeting room with kitchen. This kitchen has got only one microwave for all students. Consider that every student uses the microwave for an exponentially-distributed random time X with a two-minute mean (this is often called *the service time*).

Consider that John arrives at the kitchen and he finds that there is one person (Alice) currently warming up her food in the microwave. Thus, John must spend the following amount of time in the microwave:

$$E(R) + E(X)$$

The first term $E(R)$ refers to the average residual life of Alice's service, whereas the second term $E(X)$ is John's average service time. Essentially, Alice started warming up her food before John's arrival, so $E(R)$ refers to the average amount of time from John's arrival to the end of Alice's service time (see figure below).

However, given the fact that service times are exponentially distributed, and thanks to the memoryless nature of the exponential distribution, the past time spent by Al-

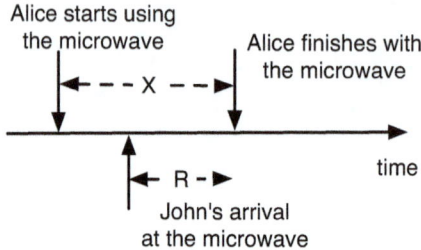

Figure 6.1: Explanation of the residual time R.

ice is not important and the average time spent by Alice in the microwave, as seen by John, is again exponentially distributed with a mean of 2 minutes. Thus:

$$E(R) = E(X) = 2 \text{ mins}$$

So, the average total time that John needs to stay in the system is:

$$E(T) = E(X) + E(X) = 4 \text{ mins}$$

in other words, 2 minutes for Alice's service time plus another 2 minutes for his own service time.

Following this reasoning, if John finds N persons in the system, he must wait $(N+1)E(X)$ on average until he exits the microwave (i.e. John waits $NE(X)$ in queue plus another $E(X)$ in the microwave).

Example 1

Now, consider that the kitchen has got two microwaves, and John finds only Alice using one of them. How long does John spend in the kitchen?

Solution

This time, John does not have to wait in queue and may use the second microwave, thus his total time in the system is:

$$E(X) = 2 \text{ mins}$$

that is, his own service time.

Example 2

Now, consider that, when John arrives at the kitchen, he finds two persons (Alice and Bob) using the two microwaves and another two persons (Claire and Diane) waiting in queue. How long does John spend in the kitchen in total?

Solution

Again, thanks to the memoryless property of the exponential distribution, the residual lives of the two persons in the microwave (Alice and Bob) are exponential random variables. The first person in queue, that is, Claire will take her microwave as soon as either Alice or Bob finishes with theirs. This is the minimum of two exponential random variables with a mean of 2 minutes, there for Claire will take her microwave after 1 minute on average. Remember that the minimum of two exponential random variables is also exponentially distributed, and its rate is the sum of the two rates. After this, Diane has to wait until either Claire or the other person in the microwave complete their service, that is, 1 minute again thanks to the memoryless property of the exponential distribution. Finally, John has to wait another 1 minute until Diane or the other person complete their service. In total, John has to wait in queue an average of $E(W_q) = 3$ minutes until he can use a microwave. Thus, the average total time spent by John is:

$$E(T) = E(W_q) + E(X) = 3 + 2 = 5 \text{ mins}$$

since he spends 2 minutes on average for his own service time in the microwave.

Alternatively, we can observe that, when the two microwaves are busy, the average output rate of them is 2 persons every 2 minutes, or 1 person per minute (2 microwaves at a rate of 1 person every 2 minutes). Because John has 4 people ahead, then he has to wait until 3 of them are out to get his microwave. This comprises 3 minutes (1 minute per person ahead) of waiting time in queue $E(W_q)$.

As shown, having more microwaves reduces the average time that John waits in queue, but not his own service time.

In addition, we observe that the average time spent by John in the system directly depends on the number of people N he finds in the system upon his arrival. This number depends on two factors: the arrival rate λ of persons at the microwaves, and the average service time of each person in the microwave $E(X)$. In the following we shall use μ as the *service rate*, or the maximum rate at which customers can be dispatched per microwave:

$$\mu = \frac{1}{E(X)} \text{ person/s}$$

The next sections provide several tools to evaluate queueing systems, like John's kitchen, and other more sophisticated ones.

6.2 Definition and Kendall's notation

Definition A queue is a waiting line, just like in supermarkets and banks.

In the context of this book, we will mainly focus on data packets waiting in a queue before being processed by an interface card in a router, for example.

Queueing systems are often described using Kendall's notation, which uses three to six letters to specify the characteristics of the queueing system:

$$A/S/c/K/N/D$$

where:

- A defines the Arrival process, typically Markovian or Poisson M, but may also be deterministic D, Erlang-type E_k, generic G, etc.

- S refers to the service-time distribution, typically exponential or Markovian M, deterministic D, generic G, etc.

c denotes the number of servers, typically one or many.

K refers to the system's capacity or maximum number of customers allowed in the system (both servers and queue length). *K* could be finite or infinite. If not specified, *K* is assumed infinite.

N is the population size from which arrivals are drawn. This could be finite or infinite. If not specified *N* is assumed infinite.

D is the queueing discipline, often First Come First Served (FCFS) but could be otherwise. If not specified *D* is assumed FCFS.

Example 3

What information can you derive from the following queueing systems:

1. M/M/1
2. M/D/1
3. M/G/3/20

Solution

M/M/1 refers to a queueing system with only one server and infinite queue length. Customer inter-arrival times are exponentially-distributed with some parameter λ, and service times are also exponentially distributed with the parameter μ.

M/D/1 refers to the same queueing system as before with only one difference: service times are not random but deterministic to some fixed value.

Finally, the M/G/3/20 refers to a queueing system with three servers and space in queue for up to 17 customers. The system may have at most 20 customers in total: 3 under service and 17 waiting in queue. Customer inter-arrival times are exponentially distributed with some parameter λ, while service times follow a different probability distribution, characterized by some probability density function $f_X(t)$.

Performance metrics of a queueing system

The next figure shows a generic queueing system, along with several performance metrics that we will use in its characterization.

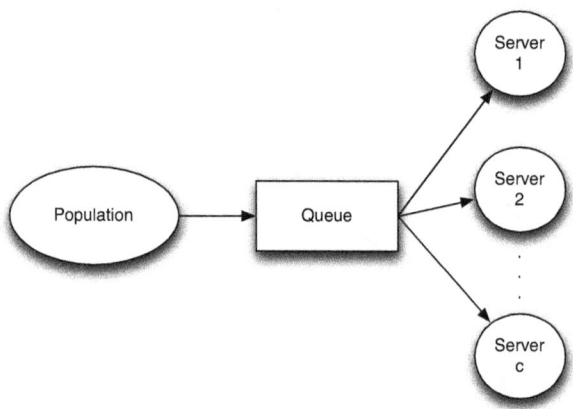

Figure 6.2: Queueing theory notation

In every queueing system, we are interested in the following set of metrics that characterize its performance:

N_q refers to the number of customers in the queue.

N_s is the number of customers under service, $0 \leq N_s \leq c$.

N denotes the total number of customers in the system, including those in the server N_s and in queue N_q, i.e.

$$N = N_q + N_s$$

X_i Service time of the i-th customer, $i = 1, 2, \ldots$. We assume that the service times X_i are independent and identically distributed random variables, characterized by some probability density function $f_X(t)$. We shall often refer to $E(X)$ as the average service time of a customer.

W_q Waiting time in a queue by a given customer.

T Total time spent in the system by a given customer, including its waiting time in queue W_q and service time X, i.e.:
$$T = W_q + X$$

Example 4

Consider an M/M/1 queueing system where service times are exponentially distributed with parameter $\mu = 10$ customer/hour. At a given time, we observe the system has 10 customers waiting in queue and another one under service. If a new customer arrives, for how long does he/she have to wait until service is provided?

Solution

First of all, the average service time of one customer in this system is:
$$E(X) = \frac{1}{\mu} = 0.1 \text{ hour}$$

Because this new customer has 10 persons ahead and another one under service, the average total time until he/she enters the server is:
$$E(W_q) = E(X_{10} + X_9 + \ldots + X_2 + X_1 + R) = 10E(X) + E(R)$$
where $E(R)$ is the average residual service time of the customer that was in the server when the new one arrived. Thanks to the memoryless property of the exponential distribution, the remaining time of this customer is also exponentially distributed with the same mean value $1/\mu$. The reader must note that
$$E(R) = E(X)$$
is true only for the exponential distribution, due to its memoryless nature.

Hence, the new customer has to wait:

$$E(W_q) = 11 \times 0.1 = 1.1 \text{ hour}$$

before entering service. After this average time, this new customer enters the server and spends an average amount of time $E(X)$ there. Hence, the total time spent by this customer in the system is:

$$E(T) = E(W_q) + E(X) = 1.1 + 0.1 = 1.2 \text{ hour}$$

The next section shows the relationship between the average number of customers in a system and the average time spent in it, thanks to *Little's theorem*.

6.3 Little's theorem

Little's theorem Let λ denote the average incoming rate of customers into the system (not necessarily exponential), $E(N)$ be the average number of customers in the system, and $E(T)$ the mean total time spent by a customer in the system.

Little's theorem states that the average number of customers in the system is the product of the mean arrival rate of customers and the mean time that a customer spends in the system. In other words:

$$E(N) = \lambda E(T) \qquad (6.1)$$

Little's theorem, although intuitively reasonable, is very powerful in queueing theory since it is not influenced by the arrival process, the service distribution, the number of servers in the system, or practically anything else.

Furthermore, Little's theorem can be applied to almost any system, and in particular, to subsystems within systems, as long as they are stable and non-preemptive. Little's theorem can also be applied to the queue:

$$E(N_q) = \lambda E(W_q) \qquad (6.2)$$

where $E(N_q)$ is the average number of customers in the queue, and $E(W_q)$ denotes the mean waiting time in the queue.

Example 5

Consider we have an external person sitting in a chair next to a bank. This bank also has a window that allows this person to take a look from time to time. This person observes the following facts:

- Between 10 and 11 a.m., ten customers arrived at the bank at 10.01 a.m., 10.10, 10.12, 10.14, 10.25, 10.33, 10.34, 10.35, 10.42 and at 10.56.

- Our observer looked over the window at 10.15 a.m., 10.30 and 10.45 observing 3, 3, and 6 customers respectively.

- Our observer did not record the departure times of these persons, although he saw some of them leaving the bank.

Can you find the average total time spent by a customer in the bank?

Solution

Thanks to Little's theorem, we know that:

$$E(T) = \frac{E(N)}{\lambda}$$

We can estimate λ from the number of customer arrivals observed. This is 10 customer arrivals in one hour (from 10 to 11 a.m.):

$$\hat{\lambda} = 10 \text{ customer/hour}$$

In addition, our observer estimates the average number of customers in the bank from his observations over the window as:

$$E(N) = \frac{3+3+6}{3} = 4 \text{ customers}$$

Hence, the average total time experienced by a customer is:

$$E(T) = \frac{4 \text{ customers}}{10 \text{ customer/hour}} = 0.4 \text{ hour}$$

or 24 minutes per customer.

6.4 The classical M/M/1 queue

The M/M/1 queue is the simplest queueing system, where customers arrive according to a Poisson process to a single server facility, and the time to serve each customer is exponentially distributed. In addition, the queue length is infinite, the population from which customers are drawn is also infinite and a First Come First Served discipline is also assumed.

This queueing system can be analyzed from the theory of CTMCs developed in the previous chapter. Essentially, customer inter-arrival times are exponentially distributed with rate λ customers per unit of time, and service times are also exponentially distributed with rate μ customers per unit of time.

The next figure shows the state-transition-rate diagram for this Markov chain, where the state label n refers to the number of customers in the system ($n = 0, 1, 2, \ldots$). As shown, the number of states in this chain are infinite since the queue can allocate an indefinite number of customers.

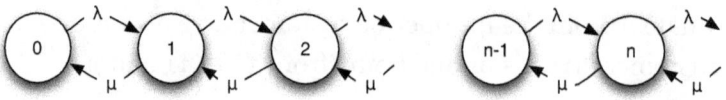

Figure 6.3: CTMC model for the classical M/M/1 system

This type of process is a particular case of the so-called *birth-and-death* process, where births (customer arrivals) occur at rate λ and deaths (customer departures) occur at rate μ.

The balance equations for the M/M/1 queue follow:

$$\lambda p_0 = \mu p_1 \Rightarrow p_1 = \frac{\lambda}{\mu} p_0$$

$$(\lambda + \mu) p_1 = \lambda p_0 + \mu p_2 \Rightarrow p_2 = \frac{\lambda}{\mu} p_1 = \left(\frac{\lambda}{\mu}\right)^2 p_0$$

$$(\lambda + \mu) p_2 = \lambda p_1 + \mu p_3 \Rightarrow p_3 = \frac{\lambda}{\mu} p_2 = \left(\frac{\lambda}{\mu}\right)^3 p_0$$

$$\vdots \qquad \vdots$$

In general, we observe the following recurrence relation:

$$p_{n+1} = \frac{\lambda}{\mu} p_n = \left(\frac{\lambda}{\mu}\right)^n p_0, \quad n = 0, 1, 2, \ldots \quad (6.3)$$

In addition, all probabilities must add up to unity:

$$\sum_{n=0}^{\infty} p_n = 1 = \sum_{n=0}^{\infty} p_0 \left(\frac{\lambda}{\mu}\right)^n \quad (6.4)$$

This summation converges if and only if:

$$\frac{\lambda}{\mu} < 1 \quad (6.5)$$

in other words, if the server is capable of dispatching customers at a higher rate μ (on average) than the customer arrival rate λ. In that case, the queue is said to be *stable*:

$$\lambda < \mu \quad (6.6)$$

In that case, the summation converges to:

$$1 = \sum_{n=0}^{\infty} p_0 \left(\frac{\lambda}{\mu}\right)^n = p_0 \frac{1}{1 - \frac{\lambda}{\mu}}$$

which yields:

$$p_0 = 1 - \frac{\lambda}{\mu} \quad (6.7)$$

Finally:

$$p_n = \left(1 - \frac{\lambda}{\mu}\right) \left(\frac{\lambda}{\mu}\right)^n, \quad n = 0, 1, \ldots \quad (6.8)$$

It is worth remarking on several facts from Eq. 6.8:

1. The probability of having exactly n customers in the system does not depend on λ or μ but on their ratio: λ/μ. This ratio is often referred to as the *system's load* or *server utilisation factor* ρ:

$$\rho = \frac{\lambda}{\mu}$$

2. The load ρ gives the percentage of time where the server is busy, that is, the system has one customer or more, i.e.:

$$\rho = \sum_{n=1}^{\infty} p_n = 1 - p_0$$

Indeed, the average number of customers on the server: $E(N_s) = \rho$ as noted from:

$$E(N_s) = 0p_0 + 1p_1 + 1p_2 + \ldots = \sum_{n=1}^{\infty} p_n = \rho \quad (6.9)$$

3. The probability density function of the number of customers in the system p_n follows a geometric distribution with parameter ρ. As such, the average number of customers and the variance in the steady state is:

$$E(N) = \sum_{n=0}^{\infty} n p_n = \frac{\rho}{1-\rho} \quad (6.10)$$

$$Var(N) = \sum_{n=0}^{\infty} (n - E(N))^2 p_n$$

$$= \frac{\rho}{(1-\rho)^2} \quad (6.11)$$

Now, we can use Little's theorem to obtain the average total time spent in the M/M/1 system as:

$$E(T) = \frac{E(N)}{\lambda} = \frac{1}{\mu} \frac{\rho}{1-\rho} = \frac{1}{\mu - \lambda} \quad (6.12)$$

Furthermore, it can be demonstrated that the total service time T in an M/M/1 queue has the following CDF:

$$F_T(t) = 1 - e^{-(\mu-\lambda)t}, \quad t \geq 0$$

that is, exponentially distributed with mean $E(T) = \frac{1}{\mu-\lambda}$.

We note that $E(T) = E(W_q) + E(X)$. Hence, the average waiting time in queue is:

$$E(W_q) = E(T) - E(X) = \frac{1}{\mu - \lambda} - \frac{1}{\mu} = \frac{\rho}{\mu - \lambda} \qquad (6.13)$$

Using Little's theorem applied to the queue only, we can obtain the average number of customers in the queue as:

$$E(N_q) = \lambda E(W_q) = \frac{\rho^2}{1 - \rho} \qquad (6.14)$$

It is a typical mistake to use:

$$E(N_q) = E(N) - 1$$

. The value 1 assumes that the server is always busy, and this is not true since $\rho < 1$.

This result can also be obtained by noting that the average number of customers in the queue is the total number of customers in the system $E(N)$ minus the average number of customers in the server $E(N_s)$:

$$E(N_q) = E(N) - E(N_s) = \frac{\rho}{1 - \rho} - \rho = \frac{\rho^2}{1 - \rho} \qquad (6.15)$$

> ### Example 6
>
> Consider an M/M/1 queue at which packets arrive following a Poisson process with rate $\lambda = 100$ packet/s and the average service time per packet is $E(X) = 2$ ms per packet. Obtain the average number of packets in the system (server and queue) and the average time spent by a packet arriving at the system. Repeat the exercise when $\lambda = 10$ packet/s and the average service time is 20 ms per packet.
>
> ### Solution
>
> First, we compute the system's load as:
>
> $$\rho = \frac{\lambda}{\mu} = \lambda E(X) = 100 \times 0.002 = 0.2$$
>
> that is, during 20% of the time, there is one packet or more in the

system. With this information, the average number of packets in the system is:

$$E(N) = \frac{\rho}{1-\rho} = \frac{0.2}{1-0.2} = 0.25 \text{ packet}$$

and the average total time spent in the system is obtained from Little's theorem as:

$$E(T) = \frac{E(N)}{\lambda} = \frac{0.25 \text{ packet}}{100 \text{ packet/s}} = 2.5 \text{ ms}$$

In the second case, we have a queue which is ten times slower than before, but the arrival rate is also ten times smaller. The system's load is then:

$$\rho = \frac{\lambda}{\mu} = \lambda E(X) = 10 \times 0.02 = 0.2$$

again 20%. The average number of packets follows:

$$E(N) = \frac{\rho}{1-\rho} = \frac{0.2}{1-0.2} = 0.25 \text{ packet}$$

same as before. However, the average time spent by a packet in the system is:

$$E(T) = \frac{E(N)}{\lambda} = \frac{0.25 \text{ packet}}{10 \text{ packet/s}} = 25 \text{ ms}$$

that is, ten times larger. This result was expected since the second server is ten times slower than the first one.

The conclusion of this exercise is very interesting: The system's load ρ and average number of packets $E(N)$ depend on the ratio between λ and μ. However, this does not apply concerning the average service and total times, $E(X)$ and $E(T)$ respectively.

Example 7

Consider a Fast Ethernet, i.e. 100 Mbps, Network Interface Card (NIC) with an infinite buffer to temporarily allocate packets. Packet arrivals are assumed to follow a Poisson process with rate $\lambda = 1000$ packet/sec. Assume that packet sizes are exponentially distributed with a mean of 625 bytes. Obtain the average total time spent by a random packet arrival at the NIC.

Solution

First, we need to compute the average packet service time:

$$E(X) = \frac{8 \cdot 625 \text{ bit/packet}}{100 \cdot 10^6 \text{ bit/s}} = 50 \mu s/\text{packet}$$

The system's load and average number of packets are:

$$\rho = \lambda E(X) = 1000 \times 50 \cdot 10^{-6} = 0.05$$

and:

$$E(N) = \frac{\rho}{1-\rho} = \frac{0.05}{1-0.05} = 0.053 \text{ packet}$$

The average time spent by a packet in the NIC is then:

$$E(T) = \frac{1}{\mu - \lambda} = \frac{1}{\frac{1}{50 \cdot 10^{-6}} - 1000} = 52.6 \mu s$$

That means, every packet experiences $50\mu s$ service time and $2.6\mu s$ of queueing delay on average.

6.5 The M/M/1/K queueing system

This queueing system is very similar to the previous M/M/1 but with a finite queue. Essentially, the system has room to allocate at most K customers, that is, 1 customer in the server and $K-1$ in the queue. If a new customer arrives and the system is full, then this customer cannot enter the queue and has to be discarded.

The next figure shows the CTMC model for the M/M/1/K queueing system. As shown, the Markov chain is very similar to the M/M/1 case but with a finite number of states: $n = 0, 1, \ldots, K$.

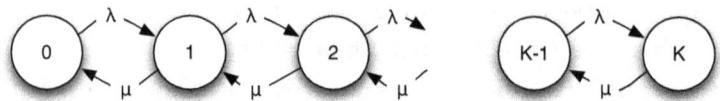

Figure 6.4: CTMC model for the M/M/1/K queueing system.

Again, solving the balance equations, we observe that:

$$p_{n+1} = \frac{\lambda}{\mu} p_n, \quad n = 0, 1, \ldots, K-1 \quad (6.16)$$

with:

$$\sum_{n=0}^{K} p_n = 1 \quad (6.17)$$

Solving, the system of equations in a similar way to that for the M/M/1 case, we obtain:

$$1 = \sum_{n=0}^{K} p_0 \left(\frac{\lambda}{\mu}\right)^n \quad (6.18)$$

It is worth noticing that this summation has a finite number of terms, thus it always converges. Therefore, it is not required to meet the stability condition of $\lambda < \mu$. The only issue is that if $\lambda > \mu$, then the system will have more arrivals than departures on average and a large number of customers will likely be rejected from the system.

Using $\rho = \frac{\lambda}{\mu}$, and solving Eq. 6.19, we have:

$$1 = \sum_{n=0}^{K} p_0 \rho^n = p_0 \frac{1 - \rho^{K+1}}{1 - \rho}, \quad \text{if } \rho \neq 1 \quad (6.19)$$

which gives the value for p_0 as:

$$p_0 = \frac{1 - \rho}{1 - \rho^{K+1}} \quad (6.20)$$

Thus, each state of the Markov chain has the following probability:

$$p_n = \frac{1 - \rho}{1 - \rho^{K+1}} \rho^n, \quad n = 0, 1, \ldots K \quad (6.21)$$

The probability that customers are rejected due to a full queue equals to the probability of finding the CTMC in state K:

$$p_K = \frac{1-\rho}{1-\rho^{K+1}}\rho^K \quad (6.22)$$

Assuming that customers arrive following a Poisson process with a rate λ, then customers are rejected at rate λp_K. Indeed, the customers who are accepted into the system are only a portion of the total. The emph acceptance rate λ_a of packets follow:

$$\lambda_a = \lambda(1-p_K) = \lambda\left(1 - \frac{1-\rho}{1-\rho^{K+1}}\rho^K\right) = \lambda\frac{1-\rho^K}{1-\rho^{K+1}} \quad (6.23)$$

and λ is referred to as the *offered* rate. The next figure shows a schematic view of the offered and acceptance rates.

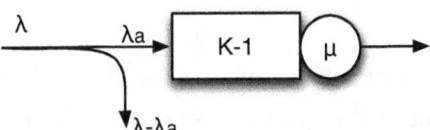

Figure 6.5: M/M/1/K queueing system. Offered acceptance rates.

The average number of customers in this queue can be obtained after solving:

$$E(N) = \sum_{n=0}^{K} np_n = \frac{\rho}{1-\rho} - \frac{(K+1)\rho^{K+1}}{1-\rho^{K+1}} \quad (6.24)$$

which approaches $\frac{\rho}{1-\rho}$ when ρ is small and K large. That is, in such cases, the M/M/1/K can be approximated to the M/M/1 queue provided that the customer rejected probability p_K is sufficiently small.

We can now apply Little's theorem to derive $E(T)$:

$$E(T) = \frac{E(N)}{\lambda_a} \quad (6.25)$$

It is worth noting that we must use λ_a since this is the actual rate of customers entering the system.

Finally:

$$E(W_q) = E(T) - E(X) = E(T) - \frac{1}{\mu} \qquad (6.26)$$

$$E(N_q) = \lambda_a E(W_q) = \lambda_a \left(\frac{E(N)}{\lambda_a} - \frac{1}{\mu} \right)$$
$$= E(N) - \rho(1 - p_K) \qquad (6.27)$$

Finally, the utilization factor or server load, that is, the percentage of time where the system has at least one customer is no longer ρ but is:

$$\sum_{n=1}^{K} p_n = 1 - p_0 = 1 - \frac{1 - \rho}{1 - \rho^{K+1}} \qquad (6.28)$$

which again approaches ρ for K large.

6.6 The M/M/c queueing system

This queueing system has a number c of servers that can be used by customers waiting in the queue. As soon as a server is free, the first customer in the queue enters this newly available server, as follows from the FCFS policy.

Again, we assume that customers arrive following a Poisson process with rate λ customers per unit of time, and customer service times are exponentially distributed with parameter μ customers per unit of time. In addition, the queue is of infinite length. The CTMC model for this queueing system is shown below.

Figure 6.6: CTMC model for the M/M/c queueing system

If only one server is busy (state 1 in the figure), then the total service rate is μ; if two servers are busy (state 2), the total service rate is 2μ, since there is a departure from the system as soon as any of the two customers complete their service. Again, the minimum of n exponential random variables with rate μ is exponentially distributed with rate $n\mu$.

In state 3, the total service rate is 3μ, and so on until all c servers are busy, yielding a total aggregated service rate of $c\mu$. At this point, new customer arrivals have to wait in queue until any server gets free.

The balance equations for this CTMC follow:

$$\lambda p_0 = \mu p_1 \Rightarrow p_1 = \frac{\lambda}{\mu} p_0$$

$$(\lambda + \mu) p_1 = \lambda p_0 + 2\mu p_2 \Rightarrow p_2 = \frac{1}{2}\frac{\lambda}{\mu} p_1 = \frac{1}{2}\left(\frac{\lambda}{\mu}\right)^2 p_0$$

$$(\lambda + 2\mu) p_2 = \lambda p_1 + 3\mu p_3 \Rightarrow p_3 = \frac{1}{3}\frac{\lambda}{\mu} p_2 = \frac{1}{3 \cdot 2}\left(\frac{\lambda}{\mu}\right)^3 p_0$$

$$\vdots = \vdots$$

In general, we observe the following recurrence relation:

$$p_{n+1} = \begin{cases} \frac{\lambda}{(n+1)\mu} p_n, & n = 0, 1, \ldots, c-1 \\ \frac{\lambda}{c\mu} p_n, & n = c, c+1, \ldots \end{cases} \quad (6.29)$$

Solving, we obtain:

$$p_n = \begin{cases} \frac{(\lambda/\mu)^n}{n!} p_0 & \text{for } n = 0, 1, \ldots, c \\ \frac{(\lambda/\mu)^n}{c! c^{n-c}} p_0 & \text{for } n \geq c \end{cases} \quad (6.30)$$

Let $\rho = \frac{\lambda}{c\mu}$ denote the *traffic intensity* of the M/M/c queueing system. Again, the chain must satisfy the condition that all p_n values must add up to unity:

$$\sum_{n=0}^{\infty} p_n = 1$$

Such a summation only converges if $\rho = \frac{\lambda}{c\mu} < 1$ as shown next:

$$1 = \sum_{n=0}^{\infty} p_n = p_0 \left(\sum_{n=0}^{c-1} \frac{(\lambda/\mu)^n}{n!} + \sum_{n=c}^{\infty} \frac{(\lambda/\mu)^n}{c! c^{n-c}} \right) \quad (6.31)$$

Now, the second summation follows:

$$\sum_{n=c}^{\infty} \frac{(\lambda/\mu)^n}{c! c^{n-c}} = \frac{(\lambda/\mu)^c}{c!} \sum_{n=c}^{\infty} \rho^{n-c} = \frac{(\lambda/\mu)^c}{c!} \frac{1}{1-\rho} \quad \text{only if } \rho < 1$$

$$(6.32)$$

Thus, eq. 6.31 reduces to:

$$p_0 = \frac{1}{\sum_{n=0}^{c-1} \frac{(\lambda/\mu)^n}{n!} + \frac{(\lambda/\mu)^c}{c!} \frac{1}{1-\rho}}, \quad \rho < 1 \qquad (6.33)$$

Again, this result is only valid if the traffic intensity is smaller than unity:

$$\rho = \frac{\lambda}{c\mu} < 1$$

which states the *stability condition* for the M/M/c queueing system. If $\rho > 1$, then the queue will grow without bounds.

A useful metric in the analysis of the M/M/c system comprises the probability that a random customer has to wait in the queue, that is:

$$p_{Wait} = p_c + p_{c+1} + p_{c+2} + \ldots = \sum_{n=c}^{\infty} p_n = p_0 \sum_{n=c}^{\infty} \frac{(\lambda/\mu)^n}{c! c^{n-c}}$$

$$= \frac{\frac{(\lambda/\mu)^c}{c!} \frac{1}{1-\rho}}{\sum_{n=0}^{c-1} \frac{(\lambda/\mu)^n}{n!} + \frac{(\lambda/\mu)^c}{c!} \frac{1}{1-\rho}} \qquad (6.34)$$

Eq. 6.34 is often referred to as the Erlang-C formula after the Danish engineer A. K. Erlang:

$$E_C(\lambda/\mu, c) = \frac{\frac{(\lambda/\mu)^c}{c!} \frac{1}{1-\rho}}{\sum_{k=0}^{c-1} \frac{(\lambda/\mu)^k}{k!} + \frac{(\lambda/\mu)^c}{c!} \frac{1}{1-\rho}} \qquad (6.35)$$

This equation gives the probability that a random customer has to wait in the queue of an M/M/c system, i.e. the probability that a customer finds all servers busy.

The next important metric in the M/M/c system is the average number of customers in the system and in the queue, which can be obtained from the Erlang-C formula as:

$$E(N) = \frac{\rho}{1-\rho} E_C(\lambda/\mu, c) + c\rho \qquad (6.36)$$

and:

$$E(N_q) = \frac{\rho}{1-\rho} E_C(\lambda/\mu, c) \qquad (6.37)$$

We observe that the average number of servers occupied in an M/M/c system $E(N_s) = E(N) - E(N_q)$ equals $c\rho$. The value of ρ measures the average load per server.

Finally, we can apply Little's theorem to obtain the average waiting time in queue as:
$$E(W_q) = \frac{E(N_q)}{\lambda} = \frac{1}{\lambda}\frac{\rho}{1-\rho}E_C(\lambda/\mu,c) = \frac{1}{c\mu - \lambda}E_C(\lambda/\mu,c) \quad (6.38)$$
and the average total time spent in the system is:
$$E(T) = E(W_q) + E(X) = \frac{1}{c\mu - \lambda}E_C(\lambda/\mu,c) + \frac{1}{\mu} \quad (6.39)$$

Example 8

Consider an M/M/2 system where packets arrive following a Poisson process with rate $\lambda = 1$ packet/s, and each of the two servers dispatch packets at rate: $\mu = 1$ packet/s. Obtain its performance metrics: $E(N)$ and $E(T)$.

Solution

First, of all the values of λ/μ and load are:
$$\frac{\lambda}{\mu} = \frac{1}{1} = 1$$
$$\rho = \frac{\lambda}{c\mu} = \frac{1}{2 \cdot 1} = 0.5$$

Next, we need to compute the Erlang-C equation for this system using eq. 6.35:
$$E_C(\lambda/\mu = 1, c = 2) = \frac{\frac{1^2}{2!}\frac{1}{1-0.5}}{\frac{1^0}{0!} + \frac{1^1}{1!} + \frac{1^2}{2!}\frac{1}{1-0.5}} = \frac{1}{1+1+1} = 0.33$$

which gives the probability that a new packet arrival has to wait in the queue because all servers are busy.
Next we calculate $E(N)$ and $E(T)$ from:
$$E(N) = \frac{\rho}{1-\rho}E_C(\lambda/\mu,c) + c\rho = \frac{0.5}{1-0.5}0.33 + 2 \cdot 0.5 = 1.33 \text{ packet}$$

and, from Little's theorem:
$$E(T) = \frac{E(N)}{\lambda} = \frac{1.33}{1} = 1.33 \text{ s}$$

The average waiting time in queue is then:

$$E(W_q) = E(T) - E(X) = 1.33 - 1 = 0.33 \text{ s}$$

and the average number of packets in the queue follows:

$$E(N_q) = \lambda E(W_q) = 0.33 \text{ packet}$$

after applying Little's theorem.

The M/M/c/c queueing system

This queueing system has c servers also, but this is the maximum number of customers in the system, i.e. there is no queue. When a customer arrives at the system, he either enters a server or is rejected. The c servers are the only resources available.

The M/M/c/c model is very useful to model calls arriving at a telephone switchboard, which usually has a finite number of channels and no waiting room. Call arrivals when all c servers are busy must be then dropped. The CTMC model for this queueing system is shown next.

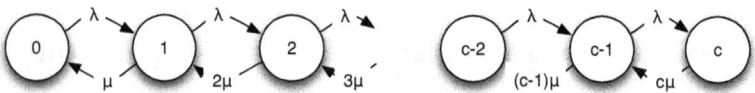

Figure 6.7: CTMC model for the M/M/c/c queueing system.

Solving the balance equations follows a similar procedure as in the M/M/c system, yielding:

$$p_n = \frac{\frac{(\lambda/\mu)^n}{n!}}{\sum_{k=0}^{c} \frac{(\lambda/\mu)^k}{k!}}, \quad 0 \leq n \leq c \quad (6.40)$$

Again, we define $\rho = \frac{\lambda}{c\mu}$ as before. However, it is worth noting that there is no need for a stability condition of type $\rho < 1$ in this case since there is no queue and the CTMC has a finite number of states. Customer arrivals who find all servers busy are simply rejected. This occurs with probability:

$$p_c = \frac{\frac{(\lambda/\mu)^c}{c!}}{\sum_{k=0}^{c} \frac{(\lambda/\mu)^k}{k!}} \quad (6.41)$$

This equation is also known as the Erlang-B loss formula:

$$E_B(\lambda/\mu, c) = \frac{\frac{(\lambda/\mu)^c}{c!}}{\sum_{k=0}^{c} \frac{(\lambda/\mu)^k}{k!}} \quad (6.42)$$

Similarly to the M/M/1/K system, the actual rate of customers that manage to enter any server is:

$$\lambda_a = \lambda(1 - p_c) \quad (6.43)$$

The mean number of customers in the system can be shown as follows:

$$E(N) = \sum_{n=0}^{c} n p_n = \frac{\lambda}{\mu}(1 - p_c) \quad (6.44)$$

Obviously, $E(N_q)$ and $E(W_q)$ are both equal to zero since there is no queue. Following Little's theorem, the average time spent by a random user that enters the system is:

$$E(T) = \frac{E(N)}{\lambda_a} = \frac{1}{\mu} \quad (6.45)$$

which equals to the average service time $E(X)$ as expected.

Example 9

Consider an M/M/3/3 system where packets arrive following a Poisson process with rate $\lambda = 3$ packet/s, and each of the three servers dispatch packets at rate: $\mu = 4$ packet/s. Obtain its performance metrics: $E(N)$ and $E(T)$.

Solution

Before computing the Erlang-B equation we need:

$$\frac{\lambda}{\mu} = \frac{3}{4} = 0.75$$

$$\rho = \frac{\lambda}{c\mu} = \frac{3}{3 \cdot 4} = 0.25$$

With these values, the probability of packet rejection due to system's full follows:

$$p_c = E_B(\lambda/\mu = 0.75, c = 3)$$

$$= \frac{\frac{0.75^3}{3!}}{\frac{0.75^0}{0!} + \frac{0.75^1}{1!} + \frac{0.75^2}{2!} + \frac{0.75^3}{3!}} = 0.033$$

The actual acceptance rate into the system is:

$$\lambda_a = \lambda(1 - p_c) = 3(1 - 0.033) = 2.9 \text{ packet/s}$$

Next we may obtain $E(N)$ and $E(T)$ from:

$$E(N) = 0p_0 + 1p_1 + 2p_2 + 3p_3$$

$$= \frac{0 \cdot \frac{0.75^0}{0!} + 1 \cdot \frac{0.75^1}{1!} + 2 \cdot \frac{0.75^2}{2!} + 3 \cdot \frac{0.75^3}{3!}}{\frac{0.75^0}{0!} + \frac{0.75^1}{1!} + \frac{0.75^2}{2!} + \frac{0.75^3}{3!}}$$

$$= 0.72 \text{ packet}$$

and:

$$E(T) = E(X) = 0.25 \text{ s}$$

6.7 Other non-classical Markovian queueing systems

In some cases, we may have a situation where customers arrive following a Poisson process and service times are also exponentially distributed, but the actual queueing system does not fit into any of the classical models studied in this chapter. In such a case, the procedure is the following:

1. Formulate the CTMC model and solve the balance equations to obtain p_n for all states. Identify p_{reject} if the system has a finite number of states. In such a case, obtain

the actual arrival rate into the system as:

$$\lambda_a = \lambda(1 - p_{reject})$$

2. Obtain the average number of customers in the steady-state as:

$$E(N) = \sum_n n p_n$$

3. Use Little's theorem to obtain the average time $E(T)$ spent by a customer that enters the system as:

$$E(T) = \frac{E(N)}{\lambda_a}$$

4. Finally, derive the average waiting time in queue $E(W_q)$ and average number of customers in the queue $E(N_q)$ as:

$$E(W_q) = E(T) - \frac{1}{\mu}, \quad E(N_q) = \lambda_a E(W_q)$$

The following example shows how to apply the above procedure.

> **Example 10**
>
> Consider a queueing system similar to an M/M/2/2 but with two different servers: a fast one capable of dispatching packets at a rate of 5 packet/s, and a slow one with a speed of 2 packet/s. We assume that packet arrivals follow a Poisson process with a rate of 1 packet/s. Packet arrivals are forwarded to the fast server if this one is available, or to the slow server otherwise. Those packets that do not make it into any server are dropped since there is no space in queue. Obtain the system's utilization and average time spent by a random customer arrival in the system (assuming it makes it to any server).
>
> Also, compute the average consumption of this queueing system under the assumption that the fast server spends 5 Watt of power, whereas the slow server

spends 2 watts on average.

Solution

This system is similar to an M/M/2/2. However, it must be noted that the two servers have different service speeds, therefore the CTMC of a conventional M/M/c/c is not valid, and a new CTMC formulation is necessary. The next figure shows a customized CTMC model for this problem. Here, state 0 represents both servers as empty, states F and S denote the situation where either the fast or the slow server are busy respectively, and state FS represents both servers as busy. In addition, λ, μ_F, and μ_S denote the packet arrival and fast and slow service rates respectively.

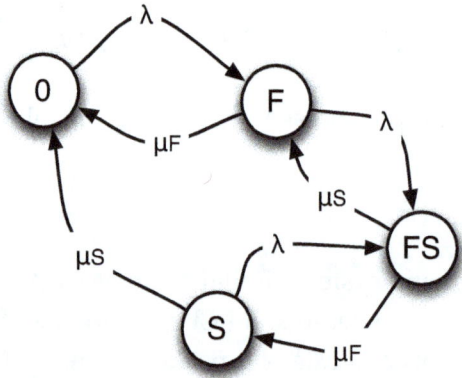

The balance equations for this CTMC are:

$$\lambda p_0 = \mu_F p_F + \mu_S p_S$$
$$(\lambda + \mu_F) p_F = \lambda p_0 + \mu_S p_{FS}$$
$$(\lambda + \mu_S) p_S = \mu_F p_{FS}$$
$$1 = p_0 + p_F + p_S + p_{FS}$$

Solving for $\lambda = 1$, $\mu_F = 5$ and $\mu_S = 2$ packet/s, we obtain:

$$p_0 = \frac{30}{38} = 0.79, \quad p_S = \frac{5/3}{38} = 0.044$$

$$p_F = \frac{16/3}{38} = 0.14, \quad p_{FS} = \frac{1}{38} = 0.026$$

Incoming packets observe both servers full with probability:
$$p_{reject} = p_{FS} = 0.026$$
so the effective incoming rate in the system is:
$$\lambda_a = \lambda(1 - p_{reject}) = 1 \times (1 - 0.026) = 0.974 \text{ packet/s}$$
The system's utilization is the probability of having at least one server busy or both:
$$Util. = 1 - p_0 = 0.21$$
The average number of packets in the system is then:
$$E(N) = 0p_0 + 1(p_S + p_F) + 2p_{FS} = 0.24 \text{ packet}$$
and the total time spent by a packet that enters the system is:
$$E(T) = \frac{E(N)}{\lambda_a} = 0.25 \text{ s}$$
which is a value between $\frac{1}{\mu_F} = 0.2$ s and $\frac{1}{\mu_s} = 0.5$ s.
Finally, the average power consumption of this system results from weighting the percentage of time when each server is full and the power spent by each server:
$$E(Cons) = 0p_0 + 2p_s + 5p_F + (2+5)p_{FS} = 0.97 \text{ Watt}$$

6.8 Further problems

Problem 1

Consider a switch with 24 input ports, and a single output port operating at 10 Mbit/s. Each input port offers packets following a Poisson process with rate 50 packet/s. Packet lengths are exponentially distributed with a mean of 625 bytes per packet. Obtain (a) the average service time per packet, (b) the probability that a packet does not have to wait in the queue, (c) the average number of packets in the system, and the average total time spent by a random packet arrival at the switch.

Solution

The average service time for a packet is:

$$E(X) = \frac{8 \times 625 \text{ bit/packet}}{10 \cdot 10^6 \text{ bit /s}} = 0.5 \text{ ms/packet}$$

Hence $\mu = 1/E(X) = 2000$ packet/s.
The system's load is then:

$$\rho = \frac{\lambda}{\mu} = \frac{24 \times 50}{2000} = 0.6$$

So the probability that a packet finds the system empty and does not have to wait in the queue is:

$$p_0 = (1-\rho)\rho^0 = 0.4$$

The average number of packets inside the switch is then:

$$E(N) = \frac{\rho}{1-\rho} = \frac{0.6}{0.4} = 1.5 \text{ packet}$$

Finally, the average time spent by a tagged packet arrival can be obtained in two different ways: the total service time for those packets ahead in the queue plus its own service time:

$$\begin{aligned} E(T) &= E(N)E(X) + E(X) = (E(N)+1)E(X) \\ &= (1.5+1)0.5 = 1.25 \text{ ms} \end{aligned}$$

Alternatively, we can apply Little's theorem, which gives the same result as before:

$$E(T) = \frac{E(N)}{\lambda} = \frac{1.5}{24 \times 50} = 1.25 \text{ ms}$$

Problem 2

Compare the performance of an $M/M/1$ queue with the service rate $\mu_1 = 2$ packet/s against the performance of an $M/M/2$ queue where each server has

$\mu_2 = 1$ packet/s service capacity. Consider two cases in this comparison: $\lambda = 0.1$ and $\lambda = 1.8$ packet/s, and find $E(N)$ and $E(T)$.

Solution

First, the M/M/1 queue has the following performance parameters:

$$E(N) = \frac{\rho}{1-\rho}, \quad E(T) = \frac{E(N)}{\lambda}$$

which gives the following results for loads: $\rho_1 = \frac{0.1}{2} = 0.05$ and $\rho_2 = \frac{1.8}{2} = 0.9$:

$$E(N_1) = \frac{0.05}{1-0.05} = 0.05 \text{ packet} \quad E(T_1) = \frac{E(N)}{\lambda} = \frac{0.05}{0.1} = 0.5 \text{ s}$$

and:

$$E(N_2) = \frac{0.9}{1-0.9} = 9 \text{ packets} \quad E(T_2) = \frac{E(N)}{\lambda} = \frac{9}{1.8} = 5 \text{ s}$$

For the M/M/c case, we first need to compute the Erlang-C equation:

$$E_C(\lambda/\mu, c) = \frac{\frac{(\lambda/\mu)^c}{c!} \frac{1}{1-\rho}}{\sum_{k=0}^{c-1} \frac{(\lambda/\mu)^k}{k!} + \frac{(\lambda/\mu)^c}{c!} \frac{1}{1-\rho}}$$

and then, $E(N)$ and $E(T)$ as:

$$E(N) = \frac{1}{1-\rho} E_C(\lambda/\mu, c) + c\rho,$$

$$E(T) = \frac{E(N)}{\lambda}$$

for $\rho_1 = \frac{\lambda_1}{c\mu} = \frac{0.1}{2 \cdot 1} = 0.05$ and $\rho_2 = \frac{1.8}{2 \cdot 1} = 0.9$:
In the first case, this is:

$$E_C(\lambda_1/\mu, c) = \frac{\frac{0.1^2}{2!} \frac{1}{1-0.05}}{\frac{0.1^0}{0!} + \frac{0.1^1}{1!} + \frac{0.1^2}{2!} \frac{1}{1-0.05}}$$

$$= \frac{0.005 \cdot 1.053}{1 + 0.1 + 0.005 \cdot 1.053} = 0.005$$

and $E(N_1)$ and $E(T_1)$ are:

$$E(N_1) = \frac{1}{1-0.05} 0.005 + 2 \cdot 0.05 = 0.1 \text{ packet}$$

$$E(T_1) = \frac{0.1}{0.1} = 1 \text{ s}$$

In the second case, we have:

$$E_C(\lambda_2/\mu, c) = \frac{\frac{1.8^2}{2!}\frac{1}{1-0.9}}{\frac{1.8^0}{0!} + \frac{1.8^1}{1!} + \frac{1.8^2}{2!}\frac{1}{1-0.9}}$$

$$= \frac{6.48 \cdot 10}{1 + 1.8 + 6.48 \cdot 10} = 0.96$$

and $E(N_2)$ and $E(T_2)$ are:

$$E(N_2) = \frac{1}{1-0.9} 0.96 + 2 \cdot 0.9 = 10.44 \text{ packet}$$

$$E(T_2) = \frac{10.44}{1.8} = 5.8 \text{ s}$$

As observed, at low loads, the average time spent in the M/M/2 system is approximately double that of the M/M/1 case. Essentially, at low loads both systems are almost empty, then the average time spent in the system is the actual average service time $E(X)$ which, in the M/M/1 case is $1/\mu_1 = 0.5$ s while in the M/M/2 case is $1/\mu_2 = 1$ s. At higher loads, both systems are most of the time very busy, and their throughputs are the same $\mu_1 = 2$ packet/s and $2\mu_2 = 2$ packet/s. For this reason, the average time spent by a random packet arrival is very similar: 5 seconds in the first case and 5.8 secs in the first case.

In conclusion, this problem shows that it is generally better to have a single server with double capacity than two servers, especially at low loads.

Problem 3

Compare the performance of a system with a number M of M/M/1 queues of average service rate μ/M each, against a single M/M/1 queue with service rate μ. Consider that both systems receive packets following a Poisson process with rate λ packet/s.

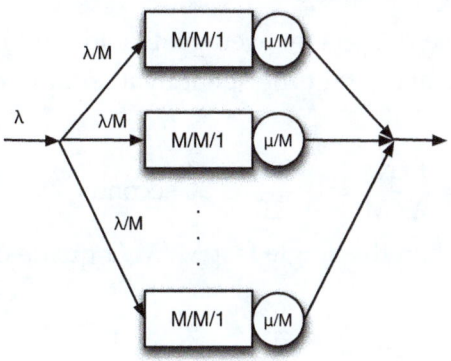

You may use $\lambda = 1$, $\mu = 1.5$ and $M = 10$ in your analysis.

Solution

Let us consider the single M/M/1 queue first. Its load is:

$$\rho = \frac{\lambda}{\mu} = 0.75$$

From theory, we know that the average number of packets in an M/M/1 queue is:

$$E(N) = \frac{\rho}{1-\rho} = \frac{0.75}{1-0.75} = 3 \text{ packets}$$

and the average total time spent by a random packet in the system is:

$$E(T) = \frac{E(N)}{\lambda} = 3 \text{ seconds}$$

In the case of several M/M/1 queues, the total traffic intensity is split into the M queueing systems, hence each system is fed with a Poisson traffic with rate λ/M. Thus, the average load per queue is:

$$\rho = \frac{\lambda/M}{\mu/M} = 0.75$$

and the total number of packets in each queue is:

$$E(N) = \frac{\rho}{1-\rho} = \frac{0.75}{1-0.75} = 3 \text{ packets per queue}$$

So each queue has three packets on average, like in the M/M/1 case. However, the average total time spent by a random packet in the system is:

$$E(T) = \frac{E(N)}{\lambda/M} = \frac{3}{1/10} = 30 \text{ seconds}$$

which is much larger than the single fast M/M/1 queue case.

Problem 4

Consider that you are the network planner of a mobile operator that needs to cover an area of 10 km^2. You have two types of Base Stations (BS): Station A with two channels only (i.e can only serve two customers simultaneously) with a cost of 10000 Euros per station and a second Base Station B that may serve up to three customers simultaneously, with a cost of 15000 Euros.

The network design requires planning for a maximum of 10 mobile phones per km^2, while each phone can be modeled with a Poisson process with a rate of 2 calls/hour in the peak hour (around 11 a.m.). The average phone call duration is 3 minutes.

Plan your network to minimize the total cost while guaranteeing that at most 5% of the phone calls are rejected.

Solution

In this design problem, we first have to identify how many base stations of type A are needed to meet the 5% rejected calls requirement, do the same for type-B base stations and compare the price of both solutions.

Let us first consider the case of type A base stations. In most telephone systems, phone calls are either accepted or rejected, but users do not wait in queue until resources are freed. Hence the most suitable queueing model in this scenario is the M/M/c/c model.

The total number of customers is $10 \cdot 10 = 100$, which yields 200 phone calls per hour aggregate, each one lasting for 3 minutes on average:

$$\lambda = 200 \text{ call/hour}$$

$$\mu = \frac{1}{E(X)} = \frac{1}{3 \text{ min/call}} \cdot \frac{60 \text{ min}}{\text{hour}} = 20 \text{ call/hour}$$

Now, consider that we partition the total area into M sub-areas or cells, each one with its base station. We further need to assume that customers are uniformly distributed across the cells, so each cell needs to cover a demand of $200/M$ phone calls. In such a case, we need to make use of the Erlang-B loss equation to find the minimum value of M that guarantees a rejection percentage smaller than 5%.

Remark that the Erlang-B loss equation follows:

$$E_B(\lambda/\mu, c) = \frac{\frac{(\lambda/\mu)^c}{c!}}{\sum_{k=0}^{c} \frac{(\lambda/\mu)^k}{k!}}$$

The value of λ/μ follows:

$$\frac{\lambda}{\mu} = \frac{200/M}{20} = \frac{10}{M}$$

In the case of A-type BS, the following results arise for different values of M cells:

$$M = 1 \Rightarrow E_B(10/1, 2) = 0.8197$$

$$M = 5 \Rightarrow E_B(10/5, 2) = 0.4000$$

$$M = 10 \Rightarrow E_B(10/10, 2) = 0.2000$$

$$M = 25 \Rightarrow E_B(10/25, 2) = 0.0541$$

$$M = 26 \Rightarrow E_B(10/26, 2) = 0.0507$$

$$M = 27 \Rightarrow E_B(10/27, 2) = 0.0477$$

So $M = 27$ type-A BSs are needed to cover this area, with a total cost of $27 \times 10k = 270k$ Euros.

For B-type BSs, we have the following results:

$$M = 1 \Rightarrow E_B(10/1, 3) = 0.7321$$
$$M = 5 \Rightarrow E_B(10/5, 3) = 0.2105$$
$$M = 10 \Rightarrow E_B(10/10, 3) = 0.0625$$
$$M = 11 \Rightarrow E_B(10/11, 3) = 0.0512$$
$$M = 12 \Rightarrow E_B(10/12, 3) = 0.0424$$

So $M = 12$ type-A base stations are needed to cover the same area, with a total cost of $12 \times 15k = 180k$ Euros, much cheaper.

Problem 5

Consider a packet switch with five input ports, a single output port, and no space in the queue. Five users are connected to this switch, each one generates traffic following a Poisson process with rate $\lambda = 10$ packet/s. Consider that the output port has either: (a) one server with service rate $\mu_a = 100$ packet/s; or (b) two servers with service rate $\mu_b = 50$ packet/s each. Compare the performance of the two possible cases (a) and (b) in terms of average delay experienced per packet and percentage of lost traffic.

Solution

Both cases (a) and (b) can be modeled with the M/M/c/c queueing system. In case (a), we have an M/M/1/1 with a total offered rate of $\lambda_{tot} = 5\lambda = 50$ packet/s and service rate $\mu_a = 100$ packet/s. Thus:

$$\frac{\lambda_{tot}}{\mu_a} = \frac{50}{100} = 0.5$$

Using the Erlang-B formula, we observe the following percentage of rejected packets:

$$p_{rej} = E_B(0.5, c=1) = \frac{\frac{0.5^1}{1!}}{1 + \frac{0.5^1}{1!}} = \frac{1}{3}$$

i.e. 33.3% packets are lost. The average delay observed for an accepted packet is:

$$E(T) = E(X) = \frac{1}{\mu_a} = 10 \text{ ms}$$

In case (b), we have an M/M/2/2 system with the same aggregated input rate, i.e. $5\lambda = 50$ packet/s, but this time the service rate is $\mu_b = 50$ packet/s per server. Thus:

$$\frac{\lambda_{tot}}{\mu_b} = \frac{50}{50} = 1$$

Using the Erlang-B formula, we observe the following percentage of rejected packets:

$$p_{rej} = E_B(1, c=2) = \frac{\frac{1^2}{2!}}{1 + \frac{1^1}{1!} + \frac{1^2}{2!}} = \frac{1}{5}$$

i.e. 20% packets are lost. The average delay observed for an accepted packet is now:

$$E(T) = E(X) = \frac{1}{\mu_b} = 20 \text{ ms}$$

So, in conclusion, the second case has a smaller packet reject ratio, since it can allocate two packets simultaneously. However, in the second case, the servers are much slower which translates into more delay than the first case.

Problem 6

In the previous scenario, consider that two computers generate high-priority packets while the other three generate low-priority packets. Consider that the output port has two servers and no space in the queue.

The first server is reserved for high-priority packets only, while the second one can be used by both high- and low-priority packets. In addition, high-priority packet arrivals first attempt to use the first server and, only if this one is busy, then they attempt to use the second server. If the second server is busy, the packet is dropped. Low-priority packets can only use the second server but, if this one is busy, then they are dropped too. Both servers operate at rate $\mu = 50$ packet/s.

Evaluate the performance of this switch in terms of delay experienced by both high- and low-priority packets.

Solution

This queueing system is similar to an M/M/2/2 system. However, the two servers operate differently, therefore the equations derived from the M/M/2/2 analysis are not valid; instead, we need to build a CTMC and solve the balance equations. To do so, we will use the following states: oo, oL, Ho, oH, HL, and HH where the first letter refers to the state of the first server (either empty or with a high-priority packet), and the second letter denotes the state of the second server (empty or full with a low- or high-priority packet). The CTMC model is shown in the figure below.

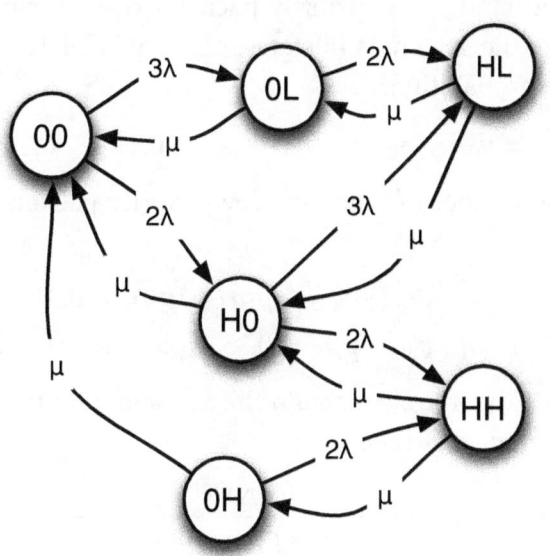

The balance equations are:

$$5\lambda p_{00} = \mu p_{0L} + \mu p_{H0} + \mu p_{0H}$$
$$(2\lambda + \mu)p_{0L} = 3\lambda p_{00} + \mu p_{HL}$$
$$(5\lambda + \mu)p_{H0} = 2\lambda p_{00} + \mu p_{HL} + \mu p_{HH}$$
$$(2\lambda + \mu)p_{0H} = \mu p_{HH}$$
$$2\mu p_{HL} = 2\lambda p_{0L} + 3\lambda p_{H0}$$
$$2\mu p_{HH} = 2\lambda p_{H0} + 2\lambda p_{0H}$$

(6.46)

together with the normalization equation:

$$p_{00} + p_{0L} + p_{H0} + p_{0H} + p_{HL} + p_{HH} = 1$$

Solving the balance balance equations brings:

$$p_{00} = 0.4337 \quad p_{0L} = 0.2551 \quad p_{H0} = 0.1531$$
$$p_{0H} = 0.0255 \quad p_{HL} = 0.0969 \quad p_{HH} = 0.0357$$

A high-priority packet arrival is rejected if it finds the system in states HL or HH. This occurs with probability:

$$p_{rej}^{(hp)} = p_{HL} + p_{HH} = 0.1326 \quad (13.26\%)$$

On the other hand, a low-priority packet arrival is rejected if it arrives when the system is in states: oL or oH or HL or HH. this occurs with probability:

$$p_{rej}^{(lp)} = p_{0H} + p_{0L} + p_{HL} + p_{HH} = 0.4132 \quad (41.32\%)$$

The average number of high- and low-priority packets in the system are:

$$E(N^{(hp)}) = p_{H0} + p_{0H} + p_{HL} + 2p_{HH} = 0.3469 \text{ packet}$$
$$E(N^{(lp)}) = p_{0L} + p_{HL} = 0.3520 \text{ packet}$$

Using Little's theorem, we obtain the average total time experienced by each packet type:

$$E(T^{(hp)}) = \frac{E(N^{(hp)})}{2\lambda(p_{00} + p_{0L} + p_{0H} + p_{H0})} = 20 \text{ ms}$$

$$E(T^{(lp)}) = \frac{E(N^{(lp)})}{3\lambda(p_{00} + p_{H0})} = 20 \text{ ms}$$

It is worth noting that, when using Little's theorem, the value of λ_{accept} in the equations refers to the average acceptance rate into the system, both for low- and high-priority packets. In addition, it is also worth noting that the total time $E(T) = E(X) = \frac{1}{\mu} = 20$ ms, as expected, since there is no waiting time in queue and both servers operate at the same service rate.

Problem 7

Consider a queueing system with two servers: one fast and one slow, with service dispatching rates of 5 and 2 packet/s, respectively. (1) Consider that packet arrivals are forwarded to the fast server if this one is available and obtain the average number of packets in the system and the average total time spent by a random packet. (2) Repeat the exercise when the policy used is that packets first try to the slow server, and only go to the fast server if the slow one is busy.

Assume that there is space in the queue for at most one packet and that packet arrivals follow a Poisson process with rate $\lambda = 1$ packet/s.

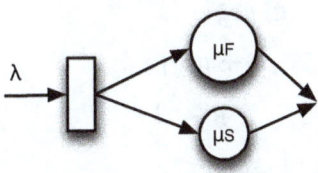

Solution

This system is similar to an M/M/2/3 queueing system with the only difference being that the two servers have different service rates. Hence, we need to build the CTMC model from scratch, solve the balance equations, and apply Little's theorem. The states are O, F, S, FS, FSQ depending on which subsystems are busy: O stands for everything empty, F means fast server busy, S means slow server busy, FS refers to both fast and slow servers busy, and finally FSQ means that both servers and the queue are full. The following figure shows the CTMC for this problem:

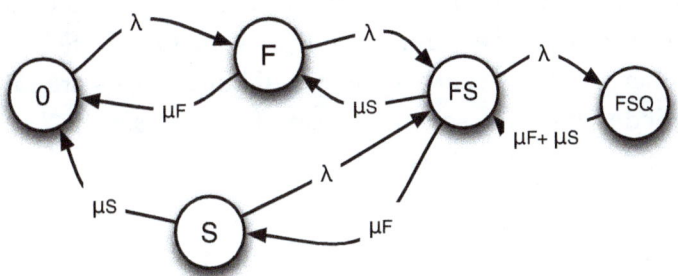

The balance equations are:

$$\begin{align}
\lambda p_0 &= \mu_F p_F + \mu_S p_S \\
(\lambda + \mu_F) p_F &= \lambda p_0 + \mu_S p_{FS} \\
(\lambda + \mu_S) p_S &= \mu_F p_{FS} \\
\lambda p_{FS} &= (\mu_S + \mu_F) p_{FSQ} \\
1 &= p_0 + p_S + p_F + p_{FS} + p_{FSQ}
\end{align}$$

The infinitesimal generator Q is then:

$$Q = \begin{pmatrix}
-\lambda & \lambda & 0 & 0 & 0 \\
\mu_F & -(\lambda + \mu_F) & 0 & \lambda & 0 \\
\mu_S & 0 & -(\lambda + \mu_S) & 0 & \lambda \\
0 & \mu_S & \mu_F & -(\lambda + \mu_S + \mu_F) & \lambda \\
0 & 0 & 0 & \mu_s + \mu_F & -(\mu_S + \mu_F)
\end{pmatrix}$$

Solving the balance equations we obtain:

$$p_0 = 0.7865, \quad p_F = 0.1398, \quad p_S = 0.0437$$

$$p_{FS} = 0.0262, \quad p_{FSQ} = 0.0037$$

The average number of packets in the system is then:

$$E(N) = 0 p_0 + 1(p_F + p_S) + 2 p_{FS} + 3 p_{FSQ} = 0.2472 \text{ packet}$$

and the average total time spent by a packet is then:

$$E(T) = \frac{E(N)}{\lambda(1 - p_{FSQ})} = \frac{0.2472}{1 \cdot (1 - 0.0037)} = 0.2481 \text{ second}$$

In the second case, the slow server is given priority over the fast server, which requires reformulating the CTMC. The only difference is that the transition rate λ from state O to state F does not exist, instead, we have a transition from state O to state S with rate λ. In other words, packet arrivals when both servers are idle are forwarded to the slow server, instead of the fast server. The new infinitesimal generator Q is then:

Solving the balance equations, we obtain the following limiting-state probabilities:

$$p_0 = 0.6481, \quad p_F = 0.0144, \quad p_S = 0.2881$$

$$p_{FS} = 0.0432, \quad p_{FSQ} = 0.0062$$

The average number of packets in the system and the average total time spent by a random packet are:

$$E(N) = 0p_0 + 1(p_F + p_S) + 2p_{FS} + 3p_{FSQ} = 0.4074 \text{ packet}$$

$$E(T) = \frac{E(N)}{\lambda(1 - p_{FSQ})} = \frac{0.4074}{1 \cdot (1 - 0.0062)} = 0.4099 \text{ second}$$

Problem 8

Consider a queueing system with three servers and no space in the queue: Two servers operate at a rate of 20 packet/s whereas the other one operates at 10 packet/s. A packet arrival will be directed to the fast servers unless these two are busy. In such a case, the packet goes to the slow server. If this server is busy too, then the packet is dropped. Consider that packets arrive at the switch following a Poisson process with rate 25 packet/s. Obtain (1) The state-transition-rate diagram for this queueing system, specifying the meaning of the states and the transition rates between them; and (2) the average number of packets in the system, and the average time spent by a random packet in the system.

Solution

The figure below shows the CTMC model for this queueing system:

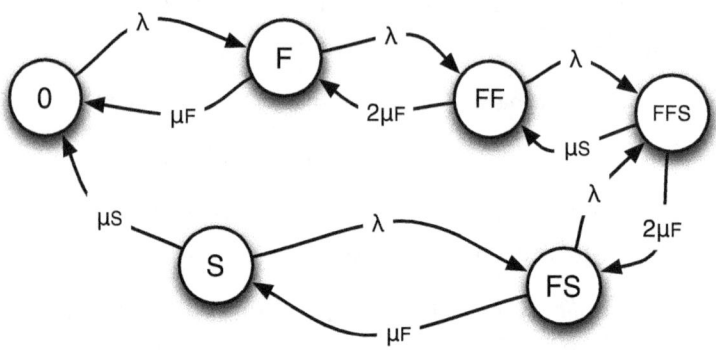

As shown, six states are necessary: 0 (all servers empty), F (one fast server busy), FF (two fast servers busy), FFS (two fast and the slow server busy), FS (one fast and one slow server full) and S (only slow server busy).

The infinitesimal generator Q is:

$$Q = \begin{pmatrix} 0 & F & FF & FFS & S & FS \\ -\lambda & \lambda & 0 & 0 & 0 & 0 \\ \mu_F & -(\lambda+\mu_F) & \lambda & 0 & 0 & 0 \\ 0 & 2\mu_F & -(\lambda+2\mu_F) & \lambda & 0 & 0 \\ 0 & 0 & \mu_S & -(\mu_S+2\mu_F) & 0 & 2\mu_F \\ \mu_S & 0 & 0 & 0 & -(\lambda+\mu_S) & \lambda \\ 0 & 0 & 0 & \lambda & \mu_F & -(\lambda+\mu_F) \end{pmatrix}$$

Solving for the steady-state probabilities, we obtain (in the same order as matrix Q):

$$p_0 = 0.2013, \quad p_F = 0.2061, \quad p_{FF} = 0.1024$$
$$p_{FFS} = 0.1368, \quad p_S = 0.0761, \quad p_{FS} = 0.1664$$

The average number of packets in the system follows:

$$E(N) = 0p_0 + 1(p_S + p_F) + 2(p_{FF} + p_{FS}) + 3p_{FFS} = 1.23 \text{ packets}$$

And the average total time spent by a packet that enters the system is:

$$E(T) = \frac{E(N)}{\lambda(1-p_{FFS})} = \frac{1.23}{25(1-0.1368)} = 0.057 \text{ seconds}$$

which is a value between $\mu_F^{-1} = 0.05$ and $\mu_S^{-1} = 0.1$ seconds (closer to μ_F in this case).

Problem 9

Consider an M/M/1/1 system at which two types of packets may arrive: High-priority and Low-priority packets. The system allows high-priority packets to preempt low-priority ones, that is, if a high priority packet arrives while a low-priority packet is under service, then the low-priority packet is dropped and the high-priority packets enters the server.
Consider that packet arrivals follow a Poisson process with rate $\lambda = 1$ packet/s, whereas only 20% of them are of high-priority. Packet service times are also exponentially distributed with a mean of 100 ms per packet regardless of its priority.

Solution

Again a new CTMC model needs to be defined since this problem does not fit to a classical M/M/1/1 system. The figure below shows the CTMC model assuming three states: 0 (server idle), H (server occupied by a high-priority packet) and L (server occupied by a low-priority packet).

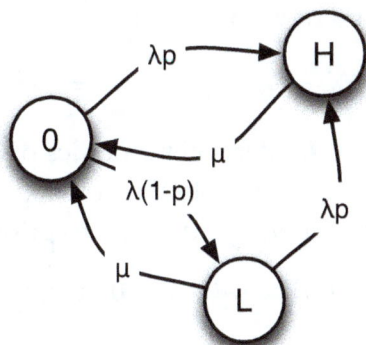

As shown, if a new high-priority packet arrives in the L state, then the low-priority packet is dropped and the high-priority one occupies its place. The infinitesimal generator Q is then:

$$Q = \begin{pmatrix} -\lambda & \lambda(1-p) & \lambda p \\ \mu & -(\lambda p + \mu) & \lambda p \\ \mu & 0 & -\mu \end{pmatrix}$$

which produces the following set of balance equations:

$$\lambda p_0 = \mu p_H + \mu p_L$$
$$\lambda(1-p)p_0 = (\lambda p + \mu)p_L$$
$$1 = p_0 + p_L + p_H$$

where $p = 0.2$, $\lambda = 1$ packet/s and $\mu = 1/E(X) = 1/0.1 = 10$ packet/s. Solving, we obtain:

$$p_0 = 0.9091, \quad p_L = 0.0178, \quad p_H = 0.0731$$

The average number of packets in the system is then:

$$E(N) = 0 p_0 + 1(p_L + p_H) = 0.0909$$

Clearly, high-priority packet arrivals are dropped when they find the system in state p_H. However, low-priority packets are dropped when they find the system in either state L or H ($p_L + p_H$). In addition, if in state L a high-priority packet arrives, then the low-priority packet under service is also dropped. The probability that a high-priority packet arrival occurs before the low-priority packet under service departs from the server is:

$$P(X_{arrival} < X_{LPservice}) = \frac{\lambda p}{\lambda p + \mu} = \frac{0.2}{0.2 + 10} = 0.02$$

as it follows from the comparison properties of the exponential distribution. Thus, the probability that a low-priority packet enters the system and is not pre-empted by a low-priority packet is:

$$(1 - p_L - p_H) \times \frac{\mu}{\lambda p + \mu} = 0.91 \times 0.98 = 0.89$$

On the other hand, high-priority packets only require to find the server empty to actually enter the system, which occurs with probability:

$$1 - p_H = 0.93$$

All packets, either low-priority or high-priority, assuming they manage to enter the system and are not pre-empted, experience an average total time in the system of:

$$E(T) = E(X) = 0.1 \text{ second}$$

Problem 10

Consider the queueing system in the figure below.

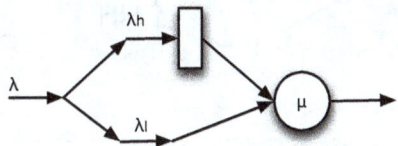

In this system, we have packet arrivals from two different classes of service: high- and low-priority. Packets are classified at ingress and forwarded along the appropriate path inside the queueing system. As observed, low-priority packets have no queue, so they either enter the server or they are dropped. On the contrary, high-priority packets have a "better chance" to receive service since they have a queue with room for one packet only.

Analyze this queueing system from the perspective of both high- and low-priority traffic. In your analysis, consider that $\lambda_l = 2$, $\lambda_h = 1$ and $\mu = 1$ packet/s.

Solution

Again, this queueing system requires a customized CTMC model with five possible states: 0 (system empty), L (server busy with a low-priority packet), H (server busy with a high-priority packet), LH (server busy with a low-priority packet and one high-priority packet in the queue) and HH (server busy with a high-priority packet and one high-priority packet in queue). The CTMC model is depicted below:

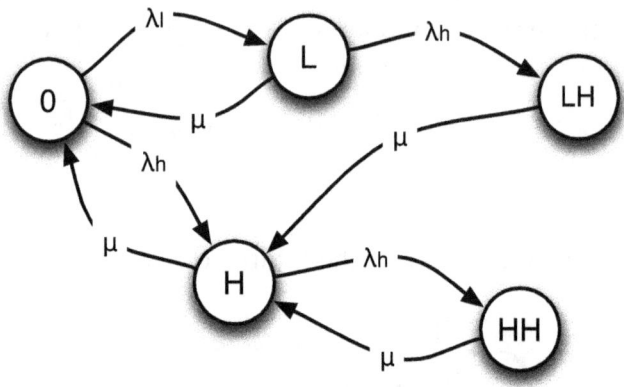

service.

The balance equations for this CTMC are:

$$(\lambda_l + \lambda_h)p_0 = \mu p_h + \mu p_l$$
$$\lambda_l p_0 = (\mu + \lambda_h)p_l$$
$$\lambda_h p_l = \mu p_{lh}$$
$$\lambda_h p_h = \mu p_{hh}$$
$$1 = p_0 + p_l + p_h + p_{lh} + p_{hh}$$

With the numbers provided, this set of equations reduces to:

$$3p_0 = p_h + p_l$$
$$2p_0 = 2p_l$$
$$p_l = p_{lh}$$
$$p_h = p_{hh}$$
$$1 = p_0 + p_l + p_h + p_{lh} + p_{hh}$$

Thus, we observe that:

$$p_0 = p_l = p_{lh}, \quad p_h = p_{hh}$$

In addition, from the first equation, we note that:

$$p_h = 3p_0 + p_l = 4p_0$$

Solving for p_0:

$$p_0 + p_l + p_h + p_{lh} + p_{hh} = p_0(1 + 1 + 4 + 1 + 4) = 1 \quad \Rightarrow$$

$$p_0 = \frac{1}{1+1+4+1+4} = \frac{1}{11} = p_l = p_{lh}$$

and:
$$p_h = p_{hh} = 4p_0 = \frac{4}{11}$$

The average number of packets $E(N)$ in the system is then:
$$E(N) = 0p_0 + 1(p_l + p_h) + 2(p_{lh} + p_{hh}) = \frac{15}{11} \text{ packet}$$

Now, low-priority packets are rejected when the server is full, which occurs with probability:
$$p_{rej,l} = p_l + p_h + p_{lh} + p_{hh} = \frac{10}{11}$$

On the other hand, high-priority packets are rejected with probability:
$$p_{rej,h} = p_{lh} + p_{hh} = \frac{5}{11}$$

As shown, high-priority packets have a better chance to make it to the server than low-priority packets, as expected. The acceptance rates for each type of packet in the system is:
$$\lambda_{a,l} = \lambda_l(1 - p_l - p_h - p_{lh} - p_{hh})$$
$$= 2 \cdot \frac{1}{11} = \frac{2}{11} \text{ low-priority packet/s}$$
$$\lambda_{a,h} = \lambda_h(1 - p_l - p_h) = 1 \cdot \frac{6}{11} = \frac{6}{11} \text{ high-priority packet/s}$$

The total acceptance rate is then:
$$\lambda_a = \lambda_{a,l} + \lambda_{a,h} = \frac{8}{11} \text{ packet/s}$$

Concerning the average time spent by each type of traffic in the system, we have that:
$$E(T_l) = \frac{E(N_l)}{\lambda_{a,l}} = \frac{1p_l + 1p_{lh}}{\lambda_{a,l}} = \frac{2/11}{2/11} = 1 \text{ sec}$$

as expected since low-priority packets only suffer the average service time $E(X) = 1$ sec. Regarding $E(T_h)$:
$$E(T_h) = \frac{E(N_h)}{\lambda_{a,h}} = \frac{1p_h + 1p_{lh} + 2p_{hh}}{\lambda_{a,h}} = \frac{13/11}{6/11} = \frac{13}{6} \text{ sec}$$

The average total service can be obtained from Little's theorem as well:
$$E(T) = \frac{E(N)}{\lambda_a} = \frac{15/11}{8/11} = \frac{15}{8} \text{ sec}$$
This result can be also obtained by weighting the average time spent by each type of traffic with its weight over the total:
$$E(T) = \frac{\lambda_{a,l}}{\lambda_a} E(T_l) + \frac{\lambda_{a,h}}{\lambda_a} E(T_h) = \frac{2/11}{8/11} \cdot 1 + \frac{6/11}{8/11} \cdot \frac{13}{6} = \frac{15}{8} \text{ sec}$$

7
Open queueing networks

7.1 Introduction

In many real situations, especially in computer and communication networks, packets departing from one queueing system enter another queueing system. For example, consider the two stations in tandem shown in the next figure.

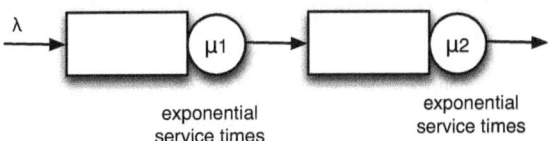

Figure 7.1: Two queues in tandem.

In this case, we have two queues connected *in tandem* (also called *feed-forward queues*), that means, the output of the first queue feeds the second queue. The two queues have infinite queue length and independent and exponentially-distributed service times with parameters μ_1 and μ_2 respectively. In addition, we know that the first queue is fed with Poisson traffic with rate λ packet/s. Hence, we can conclude that the first queue meets all the requirements of an M/M/1 queue, i.e. infinite queue length, single server, exponential inter-arrival times and exponential service times.

The first question in this simple queueing network is:

Is the second queue also M/M/1?

The second queueing system has an infinite queue length, single server, and exponential service times. Hence, the second queue is an M/M/1 if its input is a Poisson process, equivalently, if the output of the first queue is a Poisson process. Thus, concerning the output of the first queue, *are packet inter-departure times exponentially distributed?* Burke's theorem explained in the next section addresses this question.

7.2 Burke's theorem

Burke's Theorem The departure process of a stable M/M/1 queue with arrival and service rates λ and μ respectively, is a Poisson process with rate λ. In other words, customer inter-departure times are characterized by the following probability distribution:

$$F_D(t) = 1 - e^{-\lambda t}, \quad t \geq 0 \tag{7.1}$$

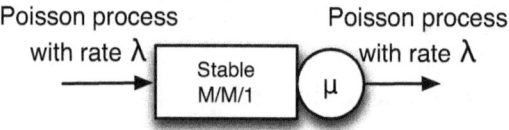

Figure 7.2: Result of the Burke's theorem.

The proof of this theorem is as follows: In a stable M/M/1 queue, we may consider two situations upon the arrival of a tagged packet: (1) the server may be empty (this occurs with probability $p_0 = 1 - \rho$) or (2) the server may be busy (which occurs with probability $1 - p_0 = \rho$).

As long as the queue is not empty (second case), packets will depart with the same inter-departure time distribution

as the service time distribution (that is, exponential with rate μ):
$$1 - e^{-\mu t} \tag{7.2}$$

When the queue is empty (first case), the inter-departure times of two consecutive packets are the sum of two random variables: the waiting time until the next arrival plus its service time. These two random variables are exponentially distributed with rates: λ and μ respectively. Therefore, the PDF of inter-departure times in the second case is the convolution of two exponential PDFs with parameters λ and μ. The CDF of such a random variable can be prove to follow:

$$1 - \frac{\mu}{\mu - \lambda} e^{-\lambda t} + \frac{\lambda}{\mu - \lambda} e^{-\mu t} \tag{7.3}$$

The two cases combined bring:

$$\begin{aligned} F_D(t) &= \rho(1 - e^{-\mu t}) + (1 - \rho)\left(1 - \frac{\mu}{\mu - \lambda} e^{-\lambda t} + \frac{\lambda}{\mu - \lambda} e^{-\mu t}\right) \\ &= 1 - e^{-\lambda t}, \quad t \geq 0 \end{aligned} \tag{7.4}$$

which is the CDF of an exponential distribution with parameter λ. In conclusion, the output of a stable queue ($\rho < 1$) fed by a Poisson process with rate λ is also a Poisson process with the same rate λ.

This conclusion is somehow reasonable: Consider a house hosting a social event. This house has two doors: front and rear. Consider that people arriving at the house use the front door, while the rear door is only used for departures. If people arrive at the house at rate λ, then they must leave exactly at the rate λ. Otherwise, if the arrival rate is larger than the departure rate in the long run, then the house must be "eating" people. On the other hand, if the arrival rate is smaller than the departure rate, then the house is "creating" new people, which does not make sense either.

In addition, the Burke's theorem states that if both inter-arrival and service times are exponentially distributed (M/M/1), then the inter-departure times are also exponentially-distributed.

7.3 Analysis of two M/M/1 queues in tandem

Consider the example of two queues in tandem proposed before. We can analyze the system using a CTMC-based model, where the states are labeled XY. Here, the first letter X denotes the number of packets in the first system, whereas Y refers to the number of packets in the second system. The figure below shows the resulting CTMC model for two stations in tandem.

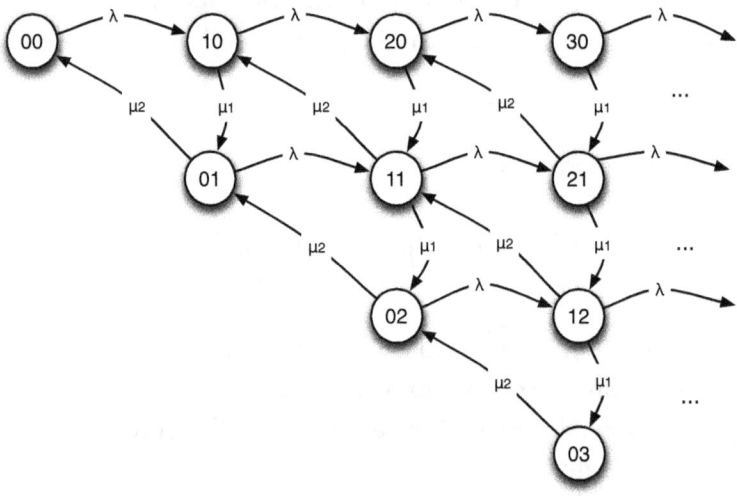

Figure 7.3: CTMC model for two M/M/1 queues in tandem.

Concerning the balance equations for this CTMC, it can be observed by induction the following relationships:

$$(\lambda + \mu_1 + \mu_2)p_{n,m} = \lambda p_{n-1,m} + \mu_1 p_{n+1,m-1} + \mu_2 p_{n,m+1}$$
$$n, m > 0$$
$$(\lambda + \mu_1)p_{n,0} = \lambda p_{n-1,0} + \mu_2 p_{n,1}, \quad n > 0$$
$$(\lambda + \mu_2)p_{0,m} = \mu_1 p_{1,m-1} + \mu_2 p_{0,m+1}, \quad n > 0$$
$$\lambda p_{00} = \mu_2 p_{01}$$
$$\sum_{m,n} p_{m,n} = 1 \tag{7.5}$$

For example, state 11 produces the following balance

equation:

$$(\lambda + \mu_1 + \mu_2)p_{11} = \lambda p_{01} + \mu_1 p_{20} + \mu_2 p_{12}$$

Solving this CTMC may require some effort. However, there is an alternative (and easier) way to solve an open queueing network, thanks to Jackson's theorem.

7.4 Types of networks and Jackson's theorem

There are two types of queueing networks:

Open networks where new packets may arrive from outside the system, and may also leave the system.

Closed networks where the number of packets in the system is fixed, and no new packet may enter or leave the system.

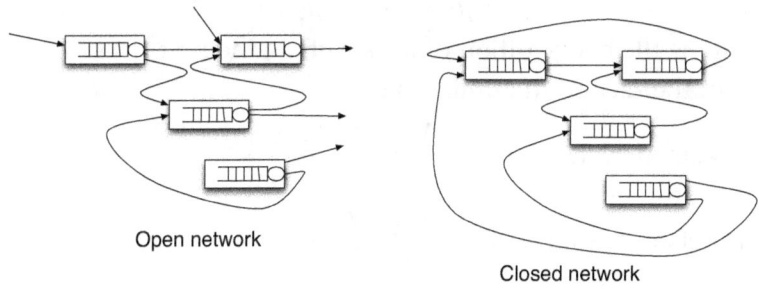

Figure 7.4: Types of queueing networks.

Open Jackson network A network of m interconnected queues is known as a Jackson network if it meets the following conditions:

1. The network is *open* and external arrivals form a Poisson process.
2. All service times are exponentially distributed and the service discipline at all queues is FCFS.

3. A customer completing service at queue i will either move to some new queue j with probability p_{ij} or leave the system with probability $1 - \sum_{j=1}^{m} p_{ij}$, which is non-zero for some subset of the queues.
4. The load ρ of all of the queues is smaller than one.

Jackson's theorem In an open Jackson network, the equilibrium state probability distribution exists for state $(k_1, k_2, ..., k_m)$ and follows a product-form solution of individual queue equilibrium distributions:

$$p(k_1, k_2, \ldots, k_m) = \prod_{i=1}^{m} p_i(k_i) = \prod_{i=1}^{m} \left[\rho_i^{k_i}(1 - \rho_i) \right] \quad (7.6)$$

For example, the case of two stations in tandem introduced at the beginning of this chapter is an open Jackson network, hence Jackson's theorem may be applied. Thus the balance equations of Eq. 7.5 reduce to:

$$p_{n,m} = \left[\left(\frac{\lambda}{\mu_1} \right)^n \left(1 - \frac{\lambda}{\mu_1} \right) \right] \left[\left(\frac{\lambda}{\mu_2} \right)^m \left(1 - \frac{\lambda}{\mu_2} \right) \right] \quad (7.7)$$

which essentially is the composition of n packets in the first and m packets in the second queuing system:

$$P(n \text{ packets in system 1}) = \left(\frac{\lambda}{\mu_1} \right)^n \left(1 - \frac{\lambda}{\mu_1} \right)$$

$$P(m \text{ packets in system 2}) = \left(\frac{\lambda}{\mu_2} \right)^m \left(1 - \frac{\lambda}{\mu_2} \right)$$

Next, the average number of packets in the two queueing systems follows:

$$\begin{aligned} E(N) &= \sum_{n,m} (n+m) p_{n,m} \\ &= \sum_n n(1-\rho_1)\rho_1^n + \sum_m m(1-\rho_2)\rho_2^m = \\ &= \frac{\rho_1}{1-\rho_1} + \frac{\rho_2}{1-\rho_2} \end{aligned} \quad (7.8)$$

which is the sum of the packets in the two M/M/1 queues separately, i.e. $E(N) = E(N_1) + E(N_2)$, where:

$$E(N_i) = \frac{\rho_i}{1 - \rho_i} \quad (7.9)$$

Using Little's theorem we observe that the average total time spent in the queueing system follows:

$$E(T) = \frac{E(N)}{\lambda} = \frac{1}{\mu_1 - \lambda} + \frac{1}{\mu_2 - \lambda} \qquad (7.10)$$

which again results in the sum of the two queueing systems separately: $E(T) = E(T_1) + E(T_2)$.

Example 1

Consider the three tandem queues in the figure below. Analyze the total time spent by a random user in the whole system, for $\lambda = 2$ packet/s, $\mu_1 = 3$, $\mu_2 = 4$ and $\mu_3 = 5$ packet/s.

Solution

Thanks to Burke's theorem, we know that both the input and output of the first queue is a Poisson process with rate λ packet/s. The same reasoning applies to the second and third queues. Thanks to Jackson's theorem, we can now analyze each M/M/1 queue separately. Thus, the load in each M/M/1 queue follows $\rho_i = \frac{\lambda_i}{\mu_i}$:

$$\rho_1 = \frac{\lambda}{\mu_1} = \frac{2}{3}, \quad \rho_2 = \frac{\lambda}{\mu_2} = \frac{1}{2}, \quad \rho_3 = \frac{\lambda}{\mu_3} = \frac{2}{5}$$

As shown, all three queues are stable. The average number of packets and total time in each M/M/1 queue follows:

$$E(N_1) = \frac{2/3}{1 - 2/3} = 2 \text{ packet} \quad \Rightarrow \quad E(T_1) = \frac{E(N_1)}{\lambda} = 1 \text{ s}$$

$$E(N_2) = \frac{1/2}{1 - 1/2} = 1 \text{ packet} \quad \Rightarrow \quad E(T_2) = \frac{E(N_2)}{\lambda} = 0.5 \text{ s}$$

$$E(N_3) = \frac{2/5}{1 - 2/5} = 0.66 \text{ packet} \quad \Rightarrow \quad E(T_3) = \frac{E(N_3)}{\lambda} = 0.33 \text{ s}$$

Thus, the average total time spent in the whole system is:
$$E(T) = E(T_1) + E(T_2) + E(T_3) = 1.83 \text{ sec}$$
Alternatively, the average total time can also be obtained from applying Little's theorem to the whole system:
$$E(T) = \frac{E(N)}{\lambda} = \frac{E(N_1) + E(N_2) + E(N_3)}{\lambda} = \frac{3.66}{2} = 1.83 \text{ sec}$$

7.5 Extended analysis for other open networks and Kleinrock's approximation

Consider a network of m queues, and let λ_j refer to the arrival rate of packets at the j-th queue:

$$\lambda_j = \lambda_{0j} + \sum_{i=1}^{m} \lambda_i p_{ij}, \quad i = 1, \ldots, m \quad (7.11)$$

where λ_{0j} gives the arrival rate of packets from the outside world to the j-th queueing system, and $\lambda_i p_{ij}$ is the rate of packets departing from the i-th queueing system that goes into the j-th queue. Essentially, part of the output packets of the i-th queue are assumed to feed other queues as long as $p_{ij} < 1$.

Thus:

$$P(n \text{ packets in server } j) = \left(\frac{\lambda_j}{\mu_j}\right)^n \left(1 - \frac{\lambda_j}{\mu_j}\right) \quad (7.12)$$

and the values of $E(N_j)$ and $E(T_j)$ for the j-th queueing system are:

$$E(N_j) = \frac{\rho_j}{1 - \rho_j} \quad (7.13)$$

$$E(T_j) = \frac{1}{\mu_j - \lambda_j} \quad (7.14)$$

The following example shows how to apply this analysis to solve an open network.

Example 2

Consider an open queueing network with two queueing systems, as depicted below. In this system, job arrivals follow a Poisson process with rate $\lambda = 2$ job/s and use the CPU for an exponentially distributed time interval with mean $1/\mu_1 = 0.1$ s. Then, jobs exit the system with probability $p_1 = 0.4$, or use the I/O resource for an exponential time with mean $1/\mu_2 = 0.2$ s. After using the I/O resource, the job must enter the CPU again. Obtain the average time spent by a random job in the whole system.

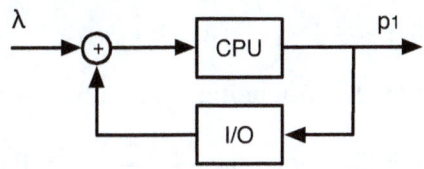

Solution

Let λ_a denote the arrival rate of packets at the input of the CPU. Thanks to Burke's theorem, the CPU outputs jobs following a Poisson process with rate λ_a job/s. In addition, only a percentage of them ($\lambda_a p_1$) exit the system, and the remaining jobs go to the I/O queue, again following a Poisson process with rate $\lambda_a(1-p_1)$. Thanks to Burke's theorem, the output of the I/O subsystem is also a Poisson process with rate $\lambda_a(1-p_1)$ and goes again to the CPU. This set of steps is summarised in the next figure:

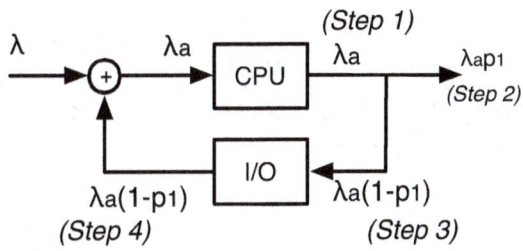

After step 4, we observe that λ_a is actually the sum of two components: λ and $\lambda_a(1 - p_1)$:

$$\lambda_a = \lambda + \lambda_a(1 - p_1)$$

solving for λ_a:

$$\lambda(1 - (1 - p_1)) = \lambda \quad \Rightarrow \quad \lambda_a = \frac{\lambda}{p_1} = \frac{2}{0.4} = 5 \text{ job/s}$$

This conclusion can also be obtained from step 2, where we observe that the output of the whole system is $\lambda_a p_1$. Obviously, if the system is stable, then the output rate must be exactly the same as the input rate, then: $\lambda_a p_1 = \lambda$.

The input rate at the I/O box is then:

$$\lambda_a(1 - p_1) = 5 \times (1 - 0.4) = 3 \text{ job/s}$$

Thus, the load of each M/M/1 queueing system follows:

$$\rho_{CPU} = \frac{\lambda_a}{\mu_{CPU}} = 0.5, \quad \rho_{I/O} = \frac{\lambda_a(1 - p_1)}{\mu_{I/O}} = 0.6$$

We observe that both queuing systems are stable (load smaller than unity), hence all assumptions are valid.

The average number of packets in each subsystem:

$$E(N_{CPU}) = \frac{\rho_{CPU}}{1 - \rho_{CPU}} = \frac{0.5}{1 - 0.5} = 1 \text{ job}$$

$$E(N_{I/O}) = \frac{\rho_{I/O}}{1 - \rho_{I/O}} = \frac{0.6}{1 - 0.6} = 1.5 \text{ job}$$

and the average total time spent in each subsystem is then:

$$E(T_{CPU}) = \frac{E(N_{CPU})}{\lambda_a} = \frac{1}{5} = 0.2 \text{ sec}$$

$$E(T_{I/O}) = \frac{E(N_{I/O})}{\lambda_a(1 - p_1)} = \frac{1.5}{3} = 0.5 \text{ sec}$$

Finally, we can apply Little's theorem to the whole system to obtain the average total time as:

$$E(T) = \frac{E(N_{CPU}) + E(N_{I/O})}{\lambda} = \frac{1 + 1.5}{2} = 1.25 \text{ sec}$$

Finally, it is worth remarking that the input process is not Poisson in open networks with feedback, such as that one in the previous example. Essentially, the input process at the CPU does not meet the requirements of independent and stationary increments demanded by a Poisson process.

However, Kleinrock's approximation states that merging several input flows on a transmission line has the effect of restoring the independence of inter-arrival times and service times. This, together with Jackson's theorem allows us to analyze open networks with feedback as several independent M/M/1 queues with good approximation results.

7.6 Further problems

Problem 1

Consider the queueing network of the figure below, with the following assumptions: (a) customer arrivals follow a Poisson process, (b) each queue has a single server, (c) service times are exponentially distributed. Compute the average load of each M/M/1 queue and the total time spent traversing the whole system. Use the auxiliar variables λ_a, λ_b, λ_c and λ_d in your analysis.

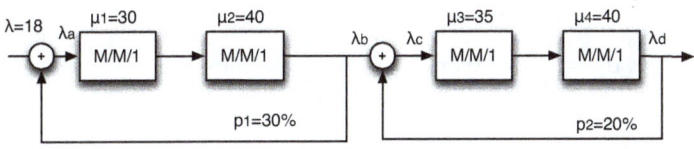

Solution

The first step is to obtain the values of the auxiliar variables: λ_a, λ_b, λ_c and λ_d in the figure.

Thanks to Burke's theorem, the output of the first queueing system is a Poisson process with rate λ_a, and so is the output of the

second queueing system. Thus, we have that:
$$\lambda_b = \lambda_a(1 - p_1)$$
and also:
$$\lambda_a = \lambda + \lambda_a p_1$$
since λ_a at the input of the first queueing system is the sum of λ and the portion of traffic coming back from the second queueing system, that is, $\lambda_a p_1$.
Similarly:
$$\lambda_d = \lambda_c(1 - p_2)$$
and
$$\lambda_c = \lambda_b + \lambda_c p_2$$

We can solve this set of equations noting that $\lambda_d = \lambda = 18$ packet/s. Hence:
$$\lambda_c = \frac{\lambda_d}{1 - p_2} = \frac{18}{1 - 0.2} = 22.5 \text{ packet/s}$$

With this value, we can obtain λ_b:
$$\lambda_b = \lambda_c(1 - p_2) = 22.5(1 - 0.2) = 18 \text{ packet/s}$$

Finally:
$$\lambda_a = \frac{\lambda_b}{1 - p_1} = \frac{18}{1 - 0.3} = 25.7 \text{ packet/s}$$

Next, we need to check whether or not the load on each M/M/1 queue is smaller than unity:
$$\rho_1 = \frac{\lambda_a}{\mu_1} = 0.86, \quad \rho_2 = \frac{\lambda_a}{\mu_2} = 0.64$$

and:
$$\rho_3 = \frac{\lambda_c}{\mu_3} = 0.64, \quad \rho_4 = \frac{\lambda_c}{\mu_4} = 0.56$$

All values of load are smaller than unity, so all queues are stable. Next, we need to obtain the average number of packets in each queueing system as $E(N_i) = \frac{\rho_i}{1-\rho_i}$:
$$E(N_1) = 6.14, \quad E(N_2) = 1.78 \text{ packets}$$

$$E(N_3) = 1.78, \quad E(N_4) = 1.27 \text{ packets}$$

In total:
$$E(N) = E(N_1) + E(N_2) + E(N_3) + E(N_4) = 10.97 \text{ packets}$$

Finally, thanks to Little's theorem, the average total time spent by a random packet is:
$$E(T) = \frac{E(N)}{\lambda} = \frac{10.97}{18} = 0.61 \text{ sec}$$

Problem 2

Obtain the average number of packets in the system and the average total time experienced by a random packet in the queueing system below. Find the values of λ_a, λ_b and λ_c in the figure.

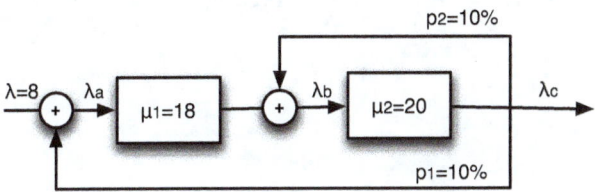

Solution

First of all, thanks to Burke's theorem, we have that the output of the first and second queues are Poisson processes with rates λ_a and λ_b respectively. Thus, from the figure, we observe that:
$$\lambda_a = \lambda + \lambda_b p_1$$

and
$$\lambda_b = \lambda_a + \lambda_b p_2$$

In addition:
$$\lambda_c = \lambda_b(1 - p_1 - p_2)$$

We know that $\lambda_c = \lambda = 8$ packet/s if the system is stable, thus:
$$\lambda_b = \frac{\lambda_c}{1 - p_1 - p_2} = \frac{8}{1 - 0.1 - 0.1} = 10 \text{ packet/s}$$

which allows to solve λ_a as:
$$\lambda_a = \lambda + \lambda_b p_1 = 8 + 10 \cdot 0.1 = 9 \text{ packet/s}$$
Thus, the average load on each queueing system is:
$$\rho_1 = \frac{\lambda_a}{\mu_1} = \frac{9}{18} = 0.5, \quad \rho_2 = \frac{\lambda_b}{\mu_2} = \frac{10}{20} = 0.5$$
The average number of packets in each M/M/1 system is:
$$E(N_1) = \frac{\rho_1}{1-\rho_1} = 1, \quad E(N_2) = \frac{\rho_2}{1-\rho_2} = 1 \text{ packet}$$
which yields:
$$E(N) = E(N_1) + E(N_2) = 2 \text{ packets}$$
Finally, thanks to Little's theorem, we can find the average total time spent in the whole queueing system:
$$E(T) = \frac{E(N)}{\lambda} = \frac{2}{8} = 0.25 \text{ sec}$$

Problem 3

Find the average time experienced by a random packet in traversing the queueing network depicted below. Use the auxiliary variables $\lambda_a, \ldots, \lambda_e$ in your analysis.

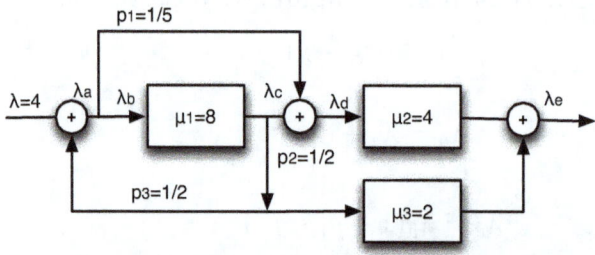

Solution

Again, we first need to find the values of the auxiliary rates $\lambda_a, \ldots, \lambda_e$. From the figure, we observe the following relationships between rates:

$$\lambda_b = \lambda_a(1 - p_1)$$

$$\lambda_a = \lambda + \lambda_b p_2 p_3$$

From these two equations, we get:

$$\lambda_a = \lambda + \lambda_a(1-p_1)p_2 p_3 \Rightarrow$$

$$\Rightarrow \lambda_a = \frac{\lambda}{1-(1-p_1)p_2 p_3} = 5 \text{ packet/s}$$

and:

$$\lambda_b = \lambda_a(1-p_1) = 5\left(1 - \frac{1}{5}\right) = 4 \text{ packet/s}$$

Next:

$$\lambda_c = \lambda_b(1-p_2) = 4\left(1 - \frac{1}{2}\right) = 2 \text{ packet/s}$$

$$\lambda_d = \lambda_c + \lambda_a p_1 = 2 + 5\frac{1}{5} = 3 \text{ packet/s}$$

$$\lambda_e = \lambda_d + \lambda_b p_2 (1-p_3) = 3 + 4\frac{1}{2}(1 - \frac{1}{2}) = 4 \text{ packet/s}$$

As shown, $\lambda_e = \lambda$ as expected.
The load values on each queueing system are:

$$\rho_1 = \frac{\lambda_b}{\mu_1} = \frac{4}{8} = \frac{1}{2}, \quad \rho_2 = \frac{\lambda_d}{\mu_2} = \frac{3}{4}, \quad \rho_3 = \frac{\lambda_b p_2 (1-p_3)}{\mu_3} = \frac{1}{2}$$

which proves that all queueing systems are stable.
Next, the average number of packets in each queueing system $E(N_i) = \frac{\rho_i}{1-\rho_i}$:

$$E(N_1) = 1, \quad E(N_2) = 3, \quad E(N_3) = 1 \text{ packets}$$

and the average time spent by a random packet in traversing the queueing network:

$$E(T) = \frac{\sum_{i=1}^{3} E(N_i)}{\lambda} = \frac{1+3+1}{4} = 1.25 \text{ sec}$$

Problem 4

Consider the network in the figure below, where a given terminal attached to node 1 generates Poisson traffic with rate $\lambda = 10000$ packets/s. All packets are destined for node 4. Packet sizes are exponentially distributed with mean 1250 bytes, and link capacities are 1 Gbit/s. Obtain the average time spent in the network if packets go from node 1 to node 4 using the following routes:

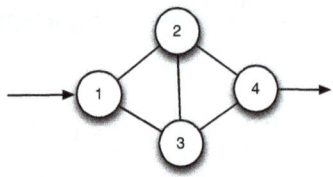

1. The routing tables force that traffic from 1 to 4 must go through 1 - 2 - 4.

2. The routing tables force that such traffic must go through 1 - 2 - 3 - 4.

3. The routing tables employ a random algorithm: At node 1, packets are forwarded to node 3 with probability $q_{13}=1/3$; then packets go from node 2 to node 3 with probability $q_{23} = 3/4$, and $q_{34} = 1$.

Solution

First, we need to construct the queuing network following the instructions provided. The resulting network in the first case is that a network with three queues in tandem:

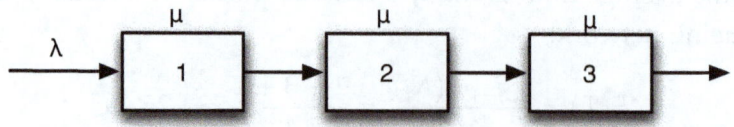

The average service time follows:
$$E(X) = \frac{8 \times 1250}{10^9} = 10\mu s$$

Since all queues have the same input rate and average service time, the load in each queue is:
$$\rho_i = \frac{\lambda_i}{\mu_i} = \frac{10000}{1/10^{-5}} = 0.1$$

resulting in the following average number of packets per queue:
$$E(N_i) = \frac{\rho_i}{1-\rho_i} = \frac{0.1}{0.9} = 0.11 \text{ packet}$$

Finally, the average total time experienced by a random packet is:
$$E(T) = \frac{E(N_1) + E(N_2) + E(N_3)}{\lambda} = 33\mu s$$

The routing rules in the second case produce a network with four queues in tandem:

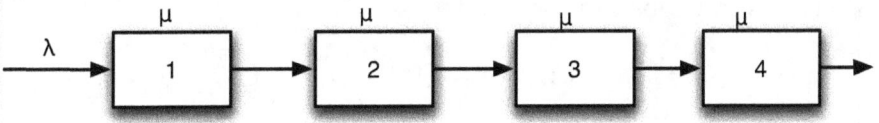

The analysis is very similar to case 1. The load in each queue is the same as before, thus the total time experienced by a random packet follows:
$$E(T) = \frac{E(N_1) + E(N_2) + E(N_3) + E(N_4)}{\lambda} = 44\mu s$$

Finally, the queueing network following the third set of instructions is the next one:

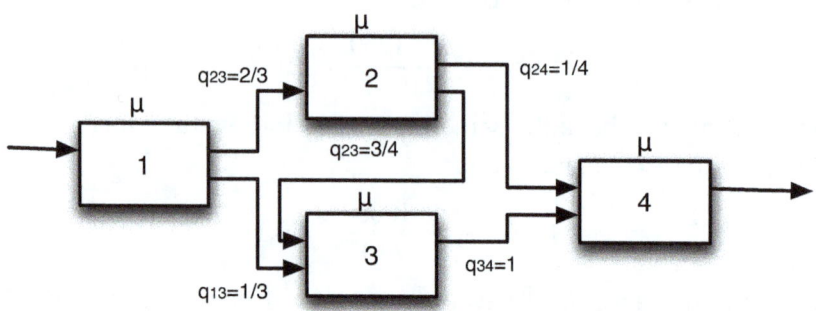

First, the value of the input rates in each system are:

$$\lambda_1 = \lambda$$
$$\lambda_2 = \frac{2}{3}\lambda_1 = \frac{2}{3}\lambda$$
$$\lambda_3 = \frac{1}{3}\lambda_1 + \frac{3}{4}\lambda_2 = \frac{1}{3}\lambda + \frac{3}{4}\frac{2}{3}\lambda = \frac{5}{6}\lambda$$
$$\lambda_4 = \frac{1}{4}\lambda_2 + \lambda_3 = \lambda$$

The average load in each system is:

$$\rho_1 = \frac{\lambda_1}{\mu} = 0.1, \quad \rho_2 = \frac{2}{3}0.1, \quad \rho_3 = \frac{5}{6}0.1, \quad \rho_4 = 0.1$$

The average number of packets in each M/M/1 system is then $E(N_i) = \frac{\rho_i}{1-\rho_i}$:

$$E(N_1) = 0.11, \quad E(N_2) = 0.07 \text{ packet}$$
$$E(N_3) = 0.091, \quad E(N_4) = 0.11 \text{ packet}$$

Finally, the average total time experienced by a random packet is then:

$$E(T) = \frac{\sum_i E(N_i)}{\lambda} = 38.1 \mu s$$

Problem 5

Obtain the average number of packets in the system and the average total time experienced by a random packet in the queueing system depicted below. Find the values of λ_a, λ_b and λ_c in the figure.

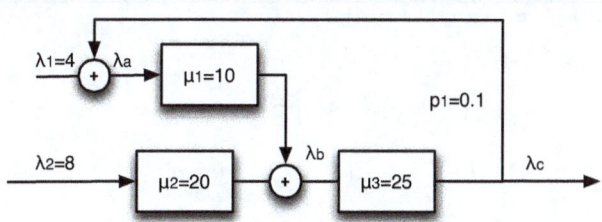

Solution

From the figure, we observe the following relationships between rates:
$$\lambda_a = \lambda_1 + \lambda_b p_1$$
$$\lambda_b = \lambda_2 + \lambda_a$$

Solving:
$$\lambda_a = \lambda_1 + (\lambda_2 + \lambda_a)p_1 \quad \Rightarrow \quad \lambda_a = \frac{\lambda_1 + \lambda_2 p_1}{1 - p_1} = 5.33 \text{ packet/s}$$

and:
$$\lambda_b = \lambda_2 + \lambda_a = 13.33 \text{ packet/s}$$

Finally, we observe that:
$$\lambda_c = (\lambda_b + \lambda_2)(1 - p_1) = 12 \text{ packet/s}$$

which is equal to the sum of λ_1 and λ_2 as expected.
The load values in each M/M/1 system:
$$\rho_1 = \frac{\lambda_a}{\mu_1} = 0.53, \quad \rho_2 = \frac{\lambda_2}{\mu_2} = 0.4, \quad \rho_3 = \frac{\lambda_b + \lambda_2}{\mu_3} = 0.67$$

and the average number of packets in the system $E(N_i) = \frac{\rho_i}{1-\rho_i}$:
$$E(N_1) = 1.13, \quad E(N_2) = 0.67, \quad E(N_3) = 2 \text{ packets}$$

The average time spent by a packet in each M/M/1 system follows $E(T_i) = \frac{E(N_i)}{\lambda_i}$:
$$E(T_1) = 0.21, \quad E(T_2) = 0.084, \quad E(T_3) = 0.15 \text{ sec}$$

Finally, the average time taken by a packet entering the network from the first port can be obtained from:

$$\begin{aligned}E(T_A) &= [E(T_1)+E(T_3)](1-p_1) \\ &+ 2[E(T_1)+E(T_3)](1-p_1)p_1 \\ &+ 3[E(T_1)+E(T_3)](1-p_1)p_1^2+\ldots \\ &= \sum_{n=0}^{\infty}(n+1)[E(T_1)+E(T_3)](1-p_1)p_1^n \\ &= [E(T_1)+E(T_3)]+[E(T_1)+E(T_3)]\frac{p_1}{1-p_1}=0.4 \text{ secs}\end{aligned}$$

which takes into account the cases where the packet traverses systems 1 and 3 only once (first item in the summation), twice (second item), etc weighted by their probability.

Following the same procedure, the packets from the second input port experience:

$$\begin{aligned}E(T_B) &= [E(T_2)+E(T_3)](1-p_1) \\ &+ [E(T_2)+E(T_3)+(E(T_1)+E(T_3))](1-p_1)p_1 \\ &+ [E(T_2)+E(T_3)+2(E(T_1)+E(T_3))](1-p_1)p_1^2+\ldots \\ &= [E(T_2)+E(T_3)](1-p_1) \\ &+ \sum_{n=1}^{\infty}[E(T_2)+E(T_3)](1-p_1)p_1^n+ \\ &+ \sum_{n=1}^{\infty}n[E(T_1)+E(T_3)](1-p_1)p_1^n \\ &= [E(T_2)+E(T_3)]+[E(T_1)+E(T_3)]\frac{p_1}{1-p_1}=0.26 \text{ secs}\end{aligned}$$

From Little's theorem, we observe that a packet selected at random experiences the following average delay:

$$E(T) = \frac{E(N)}{\lambda_1+\lambda_2} = \frac{1.13+0.67+2}{4+8}=0.31 \text{ sec}$$

We can easily check that this value matches our previous analysis:

$$E(T) = \frac{\lambda_1}{\lambda_1+\lambda_2}E(T_A)+\frac{\lambda_2}{\lambda_1+\lambda_2}E(T_B)=0.31 \text{ sec}$$

where the values of $E(T_A)$ and $E(T_B)$ are weighted by their respective probabilities (note that we have twice as many packets in the second port than in the first one.

Index

Absorbing state, 79
Acceptance rate, 143

Balance equations, 108
Bayes' theorem, 9
Binomial distribution, 5
Birth-and-death process, 136
Burke's theorem, 176

Central Limit Theorem, 10
Chapman-Kolmogorov equations, 75, 104
Complementary Cumulative Distribution Function, 3
Conditional expectation, 8
Conditional probability, 6
Continuous-Time Markov Chains, 97
Counting processes, 45
Cumulative Distribution Function, 2

Erlang-B formula, 149
Erlang-C formula, 146
Expectation, 3
Exponential distribution, 6, 27

Feed-forward queues, 175
First-passage times, 82, 111

Gamma distribution, 51
Geometric distribution, 6, 77

Holding time, 97

Independent increments, 46
Infinitesimal generator, 101
Intensity matrix, 102
Irreducible Markov chain, 79

Jackson's theorem, 179
Joint probability, 6

Kendall's notation, 130
Kleinrock's approximation, 182

Little's theorem, 134

M/M/1 queue, 136
M/M/1/K queue, 141
M/M/c queue, 144
M/M/c/c queue, 148
Markov property, 75, 104
Memorylessness, 30, 127
Minimum of random variables, 32

Normal distribution, 6

Offered rate, 143

Open Jackson network, 179

Palm-Khintchine's theorem, 54
PASTA, 125
Poisson distribution, 6
Poisson process, 47
    Aggregation of, 53
    Sampling of, 53
Probability Density Function, 2

Queue, 130
Queueing theory, 127
Queues in tandem, 175

Random variable, 1
Recurrent state, 79
Residual life, 127

Sample space, 1
Second moment, 4
Server utilisation, 138
Service rate, 130
Service time, 127
Sojourn times, 77
Stability condition, 137
State-transition diagram, 71
State-transition-rate diagram, 102
Stationary distribution, 80
Stationary increments, 46

Survival Function, 3
System's load, 138

Transition matrix, 71

Variance, 3

Total probability theorem, 9

Uniform distribution, 5

Waiting time in queue, 133

www.ingramcontent.com/pod-product-compliance
Lightning Source LLC
Chambersburg PA
CBHW050208230526
45470CB00001B/291